Conference Proceedings of the Society for Experimental Mechanics Series

Series Editor

Kristin B. Zimmerman, Ph.D.
Society for Experimental Mechanics, Inc.
Bethel, CT, USA

More information about this series at http://www.springer.com/series/8922

Yong Zhu • Alan T. Zehnder
Editors

Experimental and Applied Mechanics, Volume 4

Proceedings of the 2016 Annual Conference on Experimental and Applied Mechanics

Springer

Editors
Yong Zhu
North Carolina State University
Raleigh, NC, USA

Alan T. Zehnder
Department of Mechanical and Aerospace Engineering
Cornell University
Ithaca, NY, USA

ISSN 2191-5644 ISSN 2191-5652 (electronic)
Conference Proceedings of the Society for Experimental Mechanics Series
ISBN 978-3-319-42027-1 ISBN 978-3-319-42028-8 (eBook)
DOI 10.1007/978-3-319-42028-8

Library of Congress Control Number: 2016950284

Printed on acid-free paper

This Springer imprint is published by Springer Nature
The registered company is Springer International Publishing AG Switzerland

Preface

Experimental and Applied Mechanics represents one of ten volumes of technical papers presented at the 2016 SEM Annual Conference & Exposition on Experimental and Applied Mechanics organized by the Society for Experimental Mechanics and held in Orlando, FL, on June 6–9, 2016. The complete Proceedings also includes volumes on *Dynamic Behavior of Materials; Challenges in Mechanics of Time-Dependent Materials; Advancement of Optical Methods in Experimental Mechanics; Micro- and Nanomechanics; Mechanics of Biological Systems and Materials; Mechanics of Composite and Multifunctional Materials; Fracture, Fatigue, Failure and Damage Evolution; Residual Stress, Thermomechanics & Infrared Imaging, Hybrid Techniques and Inverse Problems; and Joining Technologies for Composites and Dissimilar Materials.*

Each collection presents early findings from experimental and computational investigations of an important topic within the field of Experimental Mechanics. This volume includes papers on: Hybrid Experimental and Computational Techniques, Advanced Experimental Mechanics Methods, Integration of Models and Experiments, Soft Materials, Education and Research in Progress, and Applications.

Raleigh, NC
Ithaca, NY

Yong Zhu
Alan T. Zehnder

Contents

Chapter 1
Effect of Specimen Holder on Static and Fatigue Tests on Titanium/Cement Interfaces

S. Riahinezhad, Y. Li, and M. Khandaker

Abstract A tension or compression load was applied onto the Ti rod to test the fracture strength and fatigue life of Ti-cement interface under static and fatigue loadings, respectively. These tests are referred as static and fatigue in this study. A customized holder for the cement is required for the static and fatigue experiments, since the typical wedge, pneumatic, or hydraulic gripper are not suitable for static and fatigue tests on the fracture tests of bi-material samples. The objectives of this study are (1) to evaluate the effect of cement thickness on the fracture strength and fatigue life on Ti-cement union by finite element analysis; (2) to evaluate the effect of plastic cement holder and aluminum cement holder on fracture strength and fatigue life on Ti-cement union by experiment and finite element analysis. Ti-cement union model with 0.22 and 0.11 in. cement, Ti-cement-holder union with plastic and aluminum holders were created and validated using ANSYS in this study to develop a suitable specimen holder for static and fatigue tests. Experimental static tests of Ti-cement with both plastic and aluminum specimen holders were conducted as well. The result clearly showed that both plastic and aluminum holders can be used for static test whereas aluminum holder required much larger fracture load compared to the fracture load on plastic holder. Plastic holder is not suitable for fatigue test, because fatigue test required a stronger and more rigid holder such as aluminum.

Keywords Titanium • Cement • Interface • PMMA • Polycaprolactone • Fracture strength • Implant

1.1 Background

The bond of an ideal implant with surrounding tissue must maintain certain fracture strength, σ_f, due to static loads (e.g. body weight, carry weight) as well as fatigue life, N, due to cyclic loads (e.g. walking, running) [1]. Accordingly setup for fracture strength and fatigue life test, especially the specimen holder, was needed in this research. To the author's best understanding, there were no static and fatigue test gripper for bi-material samples that can be suitable for Test Resources Machines (UTM). Finite element analysis on customized specimen holder for static and fatigue tests were executed in this study to validate the application of the holder for these tests.

Cement holder material is one of the main factors to consider before the cement holder design. The elastic properties of cement holder material may affect the fracture of Ti-cement interface under the static tests. If the cement holder is not strong enough, the damage will occur on the cement holder other than the Ti-cement interface. Cement thickness is also an important factor that affects Ti-cement interface. Fisher et al. [2] found out increasing cement mantle thickness caused substantial strain reductions in the distal cement which may eventually increase the fatigue life of a bone-implant system. In this study, effects of the cement holder material and cement thickness on the fracture of Ti-cement interface were investigated to determine a suitable cement holder design for the static and fatigue test of Ti-cement interface.

The objectives are (1) to evaluate the effect of cement thickness on the static and fatigue tests on Ti-cement union by finite element analysis; (2) to evaluate the effect of plastic cement holder and aluminum cement holder on static and fatigue tests on Ti-cement union by experiment and finite element analysis.

The scope of work for this study was: (1) to conduct static experiment test of Ti-cement with plastic holder and aluminum holder; (2) to determine the frictional coefficient of Ti-cement interface and cement-aluminum holder interface based on the experiment result; (3) to determine the cement thickness by comparing the deformation of Ti at Z axis, Von-Mises stress of

S. Riahinezhad • Y. Li • M. Khandaker (✉)
Department of Engineering and Physics, University of Central Oklahoma, Edmond, OK 73034, USA
e-mail: mkhandaker@uco.edu

Y. Zhu, A.T. Zehnder (eds.), *Experimental and Applied Mechanics, Volume 4*,
Conference Proceedings of the Society for Experimental Mechanics Series, DOI 10.1007/978-3-319-42028-8_1

cement and fatigue life of cement in models of Ti-cement union with 0.11 in. cement and 0.22 in. cement through ANSYS; (4) to evaluate the effect of the cement holder on the fracture of Ti-cement interface by comparing the deformation of Ti at Z axis, Von-Mises stress of cement and fatigue life of cement in models of Ti-cement-holder union with plastic and aluminum holder.

1.2 Material and Method

1.2.1 Static Test Experiment

Titanium (Ti) bars (6Al-4V ELI, ASTM B 348 standard, grade 23, biocompatible) of dimension (3″ length and 3/8″ diameter) were purchased from Titanium Metal Supply, Inc., Poway, CA. Cobalt™ HV bone cement (Biomet Inc., Warsaw, IN) [3] was used as the polymethyl methacrylate (PMMA) cement. The titanium rod is encapsulated by a 3D printed (Dimension elite 3D printer) [4] cylindrical holder or aluminum holder in order to cure the rod to the cement without being contaminated. Figure 1.1 shows these two holders for the production of titanium/cement sample. Static and fatigue tests were performed under static compression load condition using Test Resources 800LE4 universal testing machine (UTM) [5] to test the breaking load of the Ti-cement specimen under static test. During the test, Ti rod samples were fixed in the UTM. After that, the PMMA cement was prepared and poured into the bottom holder as shown in Fig. 1.2a. According to Biomet HV PMMA cement preparation protocol, PMMA cement was prepared by hand mixing 1.25 g of PMMA powder with 0.625 mL of methyl methacrylate (MMA) monomer using powder: monomer ratio of 2:1. PMMA was poured on top of the different titanium samples during doughy phase in the mold. A 60 kPa weight (clinically achievable range [6]) was placed onto the top of the cement and cured for 15 min (Fig. 1.2b). After cured, the weight was taken out and run the static test to find σ_f of Ti/PMMA samples at strain rate 0.05 mm/s. The maximum push-down force was determined and fracture strength was calculated by dividing the force at the point of failure by the surface area of the implant in contact with the cement. The surface area of the implant, A, was calculated using: $A = 2\pi rl$. where r refers to the diameter of the Ti rod which is 3/8″ and l is the height of the Ti implant surrounding with the cement, which can be obtained from the height of the cement.

1.2.2 Finite Element Analysis

Finite element 3D models of a cemented Ti rod and a cemented Ti rod with a cement holder surrounding the cement were created to study the debonding process of the stem-cement interface according to the dimension of the experimental model. The geometric FEA model was modeled and analyzed in ANSYS Workbench R15 [7]. The deformation of Ti rod at Z axis, Von-Mises stress and fatigue life of Ti-cement union with two different cement thicknesses and Ti-cement holder with two different materials were used to find the suitable cement holder thickness and material for static and fatigue tests. Deformation of Ti rod at Z axis was applied in the FEA model, which is the experimental displacement of Ti rod break from the cement in the direction normal to the Ti rod top surface. Von-Mises stress, also referred as equivalent tensile stress, is used to check whether the design will withstand a given load condition. This is accomplished by comparing the Von Mises stress to the material's yield stress. Fatigue test in this simulation was analyzed for the Ti-cement model as shown in Fig. 1.2 under static structural analysis. Fatigue life suggests the number of cycles of fluctuating stress and strain that a material will sustain before failure occurs. Modified Goodman theory was used for the prediction of life in this study. Materials used in the

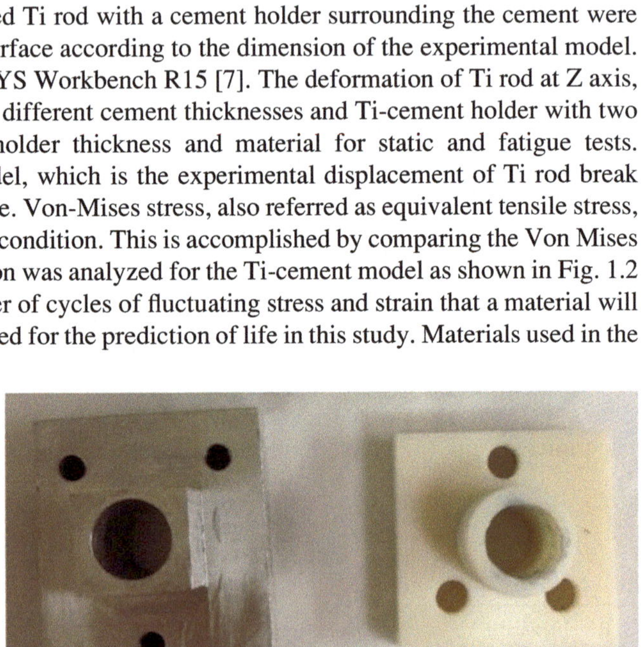

Fig. 1.1 Fabricated aluminum and plastic cement holder using CNC machining and 3D printer

Fig. 1.2 (**a**) Static tests setup in UTM before curing; (**b**) during curing

Table 1.1 Material properties used for the fatigue test of bi-material samples

Properties	Titanium alloy	PMMA cement	ABS plastic	Aluminum
Young's modulus (GPa)	116	3.3	2.2	71
Poisson's ratio	0.32	0.39	0.35	0.33
Shear modulus (GPa)	43.939	1.1871	0.81481	26.692
Density (kg/m^3)	4507	1190	1040	2770
Compressive ultimate strength (MPa)	120	120	50	310
Compressive yield strength (MPa)	970	120	65	280
Tensile ultimate strength (MPa)	950	69	40	0
Tensile yield strength (MPa)	880	70	43	280

analysis were titanium alloy (rod), PMMA cement, ABS plastic which corresponds to the 3D printer holder and aluminum alloy which corresponds to the CNC machining holder in the experimental model. Table 1.1 shows the properties of materials used for FEA analysis. A normal stiffness factor of 1 was used for the Ti-PMMA interface, and ANSYS program controlled stiffness factor was used for PMMA-ABS plastic interface.

Huiskes [8] recommended a non-uniform thickness of cement ranging from 3 to 6 mm (0.118–0.236 in.) for the proximal part of the canal. Ti-cement models with two different cement thicknesses 0.22 and 0.11 in. were created and analyzed in this study to evaluate the effect of cement thickness on the fracture of Ti-cement interface. It is assumed that there is no momentum of Ti rod and the outside surface of the cement is fully constrained. Figure 1.3a shows the meshed FE models of Ti with 0.22 in. cement with the element size of 0.8E−3 m. There were 174205 nodes and 41501 elements for Ti-cement model. For Ti-cement models, a compression load that corresponds to the failure load for the experimental model discussed in Sect. 4.2.1 was applied on the top surface of the Ti rod. Only displacement normal to the top surface was allowed for the Ti rod. The outside surface of the cement was constraint in all directions (Fig. 1.4a).

Ti-cement-holder models with two different holder materials (plastic and aluminum) were created and analyzed to evaluate the effect of cement holder on the fracture of Ti-cement interface. Figure 1.3b shows the Ti-cement-holder model with the element size of 3.0E−3 m, which has 8242.6 nodes and 8002 elements. It is also assumed that there is no momentum of Ti rod and the outside surface of the cement holder is fully constrained. Only displacement normal to the top surface was allowed for the Ti rod and the bottom surface of the cement holder was constraint in all direction (Fig. 1.4b). A compression load that corresponds to the failure load for the experimental model discussed in Sect. 1.2.1 was applied to the top surface of the titanium rod. The contact of the plastic holder and the cement was set as bonded while the contacts of aluminum-cement

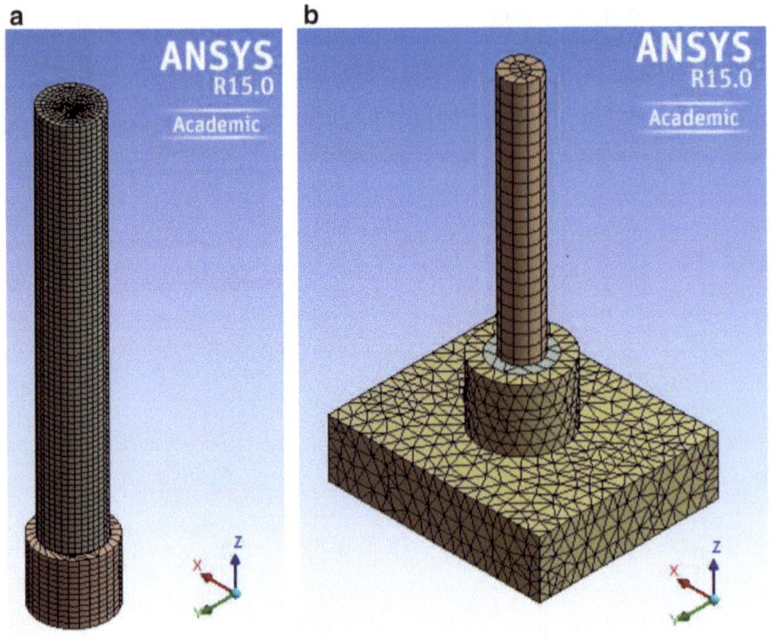

Fig. 1.3 (**a**) Meshed model of Ti-cement samples with element size of 0.8 mm; (**b**) meshed model of Ti-cement-holder samples with element size of 3 mm

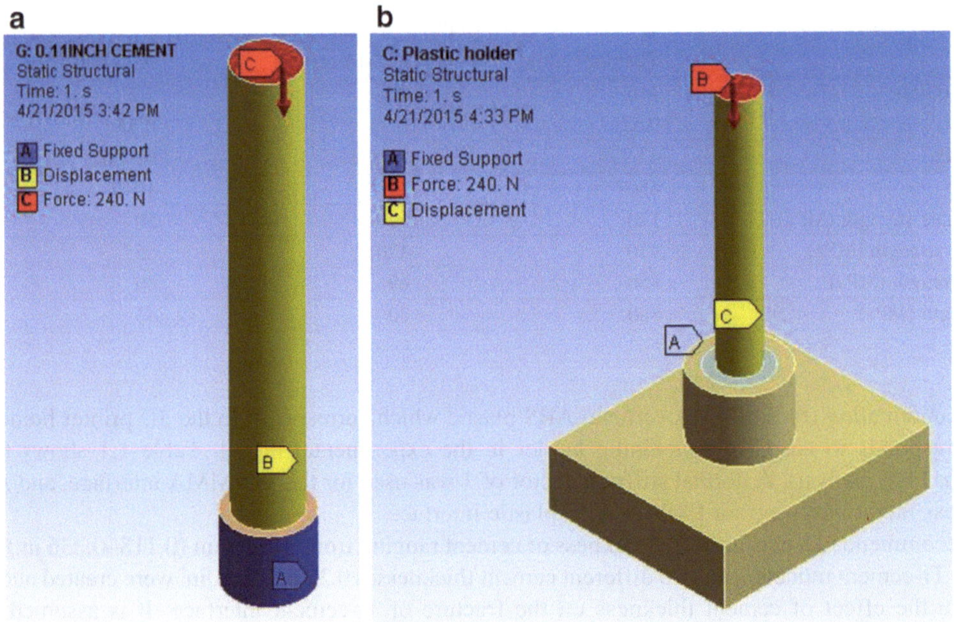

Fig. 1.4 (**a**) Boundary conditions of the Ti-0.11 in. cement model; (**b**) boundary conditions of the Ti-cement-holder model

interface and cement-Ti interface were both set as frictional contacts. The frictional coefficient would be analyzed based on the experiment result of displacement-load plot. According to displacement-load plot, the maximum failure load was applied on to the Ti rod in the model of Ti-cement and multiple frictional coefficient of Ti-cement interface would be input to find out the corresponding displacement of Ti-rod. When the displacement of Ti rod in Ti-cement union obtained from the ANSYS simulation result matched the experiment result, the input frictional coefficient was set as the frictional coefficient of Ti-cement interface and used to continue test the frictional coefficient of aluminum-cement. Similarly, the load was applied onto the Ti rod of Ti-cement-holder model and multiple frictional coefficients of aluminum-cement were input to find out the displacement of Ti rod matching the corresponding experimental result.

1.3 Results and Discussion

Control Experiment: Based on our observation of the force required to break the cement and Ti rod, the fracture load of Ti-cement interface with aluminum holder is much higher than the cement with plastic holder (Table 1.2). The most likely reason may be the difference of contact load applied from the holder to the cement. From Table 1.1, Young's modulus and shear modulus of ABS plastic is 2.2 GPa and 0.81481, while aluminum has 71 GPa for Young's Modulus and 26.692 GPa for Shear Modulus. Thus it can be seen that ABS is more elastoplastic material and when the load applied on the Ti rod transferred to cement and the plastic holder, the holder deformed more. From load-displacement result for Ti-cement interface with the aluminum holder has a higher stiffness comparing the result with the plastic holder. The other possible reason to explain the different fracture load would be the different thermal conductivities between plastic and aluminum holders. As reported by Li et al. [9], time-temperature profiles of bone cement were sensitive to the mold material. Poor conductive material, like ABS plastic, decreases the cement heat transfer generating from PMMA cement curing and have higher thermal expansion of the cement. The thermal expansion of the cement would have generated large residual stress which would affect the cement stress distributions at the stem cement interface [10]. Thus the difference of thermal expansion of the cement can be presented for different holders that may produce different the preload value during static tests. Table 1.2 shows the observation of preload during the experiment. With the plastic holder, the maximum preload before the test was from 110 to 150 N, which is around two times more than the preload of the test with the aluminum holder.

Finite element analysis: Two models of Ti-cement union with the cement thickness 0.22 and 0.11 in. were designed. Different element size of 0.8, 0.9, 1 and 2 mm were analyzed since the nodes number with element size 0.8 mm is the maximum nodes number the academic version of ANSYS R15 can go. Tables 1.3 and 1.4 show life and stress at different element size for Ti-cement and Ti-cement-holder union models. The mesh size at 0.8 mm is used due to the limitation of the ANSYS academic version. Two models of Ti-cement union with the cement thickness 0.22 and 0.11 in. were created and analyzed with different element size of 2.8, 2.9, 3 and 4 mm.

Figure 1.5 compares the fatigue life result with different meshing sizes. The mesh size at 3 mm is used since the node number with 2.8 mm mesh size is the maximum node number of ANSYS academic version.

Table 1.2 Static result of plastic and aluminum fatigue holders (n = 3)

Sample groups	Method	L (mm)	Contact area (mm^2)	Fracture load (N)	Stiffness (KN/mm)	Fracture strength (N/mm^2)	Max preload (N)
1	Plastic specimen holder	10.233 ± 0.170	307.344 ± 5.106	298.619 ± 13.350	1.679 ± 0.041	0.967 ± 0.025	0–300
2	Aluminum specimen holder	10.208 ± 0.035	306.810 ± 1.051	617.531 ± 168.483	7.893 ± 2.602	2.014 ± 0.485	0 –110

Table 1.3 Mesh sensitive of FEA model at the Ti-cement union

Element size (mm)	Element number	Node number	Life (cycles) (e9)	Alternating stress (MPa)		
				Minimum stress	Maximum stress	Mean stress
2	1274	6446	12	0.10	3.9442	2.0221
1	19,855	84,825	8.1803	0.041997	7.3591	3.7005
0.9	22,944	98,232	1.6949	0.042002	8.8166	4.4293
0.8	41,501	174,205	1.0942	0.05221	9.5146	4.7834

Table 1.4 Mesh sensitive of FEA model at Ti-cement-holder union

Element size (mm)	Element number	Node number	Life (cycles)	Alternating stress (MPa)		
				Minimum stress	Maximum stress	Mean stress
4	7490	3793	10,000	6.5206e−16	46.569	23.2845
3	8002	8242.6	8242.6	0.0005967	8.9579	4.4792
2.9	8486	16,557	8071.5	0.00062356	8.1314	4.0660
2.8	9360	17,733	8869.7	0.00025352	7.5824	3.7913

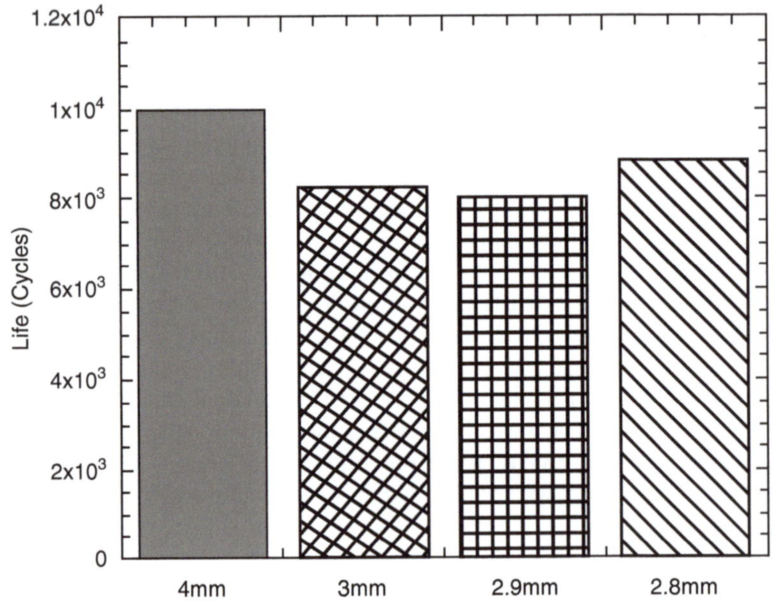

Fig. 1.5 Ti-cement-holder union fatigue life variations with different meshing element sizes

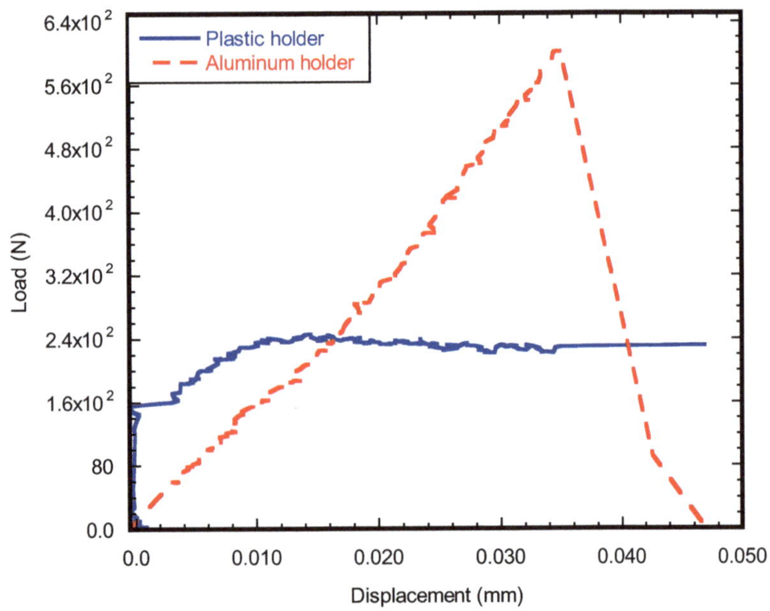

Fig. 1.6 Load vs. displacement plot for plastic and aluminum cement holder

Frictional coefficient for stem-cement and holder-cement interfaces: The frictional coefficients of Ti-cement and cement-aluminum were calculated based on the experiment load-displacement graph (Fig. 1.6). The failure load of 241.14 N has corresponding displacement 0.03002 mm for plastic holder. In finite element analysis, the load of 241.14 N was applied onto the Ti rod and frictional coefficients from 0 to 10 was tried to find the displacement of 0.03 mm in Z axis. Our FEA found that frictional coefficient 7 is the best fit for the coefficient at Ti-cement because the Z-displacement of it 0.029703 mm is the closest to the experimental displacement 0.3002. The frictional coefficient 7 was used in the continuing the frictional coefficient analysis of the aluminum holder and cement. Also, Ti-cement-aluminum holder was analyzed under the load similar to the load applied for the model of Ti-cement-plastic holder. From the original data recorded and the experiment load-displacement graph for aluminum holder (Fig. 1.6), under 240 N load the displacement of the cement from Ti was 0.016444 mm. In the same way, the load of 240 N was applied onto the Ti rod and a frictional coefficient from 10 to 50 was tried to find the deformation closest to 0.016444 mm in Z axis. Our FEA found that frictional coefficient 14 at the holder-cement because the Z-displacement of it 0.01649 is the closest to the experimental displacement 0.016444.

FEA static and fatigue results for cement with different thickness: Since the objective of the FEA was to evaluate the effect of the cement thickness on the fracture performance of Ti-cement interface, constrains applied on the Ti-cement model should be the same for different cement thickness. A downward load of 240 N that was closest to the minimum experimental failure load of the Ti-cement-holder model was applied to the top surface of the titanium rod. The frictional coefficients of Ti-cement were input as 7 and the static and fatigue simulations were performed. For the 0.22 in. cement, the Ti rod slipped from the cement 0.008 mm under the maximum Von-Misses stress of 8.8166 MPa and minimum Von-Misses stress of 0.042002 MPa (Fig. 1.7a). The Ti rod has displacement of 0.0091945 mm at Z-axis from the 0.11 in. cement, 0.001 mm more than result of the thicker cement. The Von-Mises stress of the 0.11 in. cement has the maximum value of 13.719 MPa and the minimum value of 0.029224 MPa (Fig. 1.7b). The minimum fatigue life of the 0.22 in. cement is 1.6949e9 while the life of 0.11 in. cement is 2.5623e8 (Fig. 1.8). Thus the Ti-cement with less cement would fracture faster, which is actually more suitable for observation in the experiment. However, if the cement is too thin, more loads would be transferred to the cement holder. In this case, the impact of the cement holder on the fracture of implant-cement interface will become stronger. Furthermore, a thin cement holder is also a big challenge for the experimental machining. Thus the 0.22 in. cement model was used to continue the analysis in the following analysis.

FEA static and fatigue results for plastic and aluminum holders: Based on the static experiment setup and the result, the bottom surface of the cement holder model was constraint in all direction (Fig. 1.4b). A downward load of 240 N that corresponds to the failure load for the experimental model was applied to the top surface of the titanium rod. The frictional

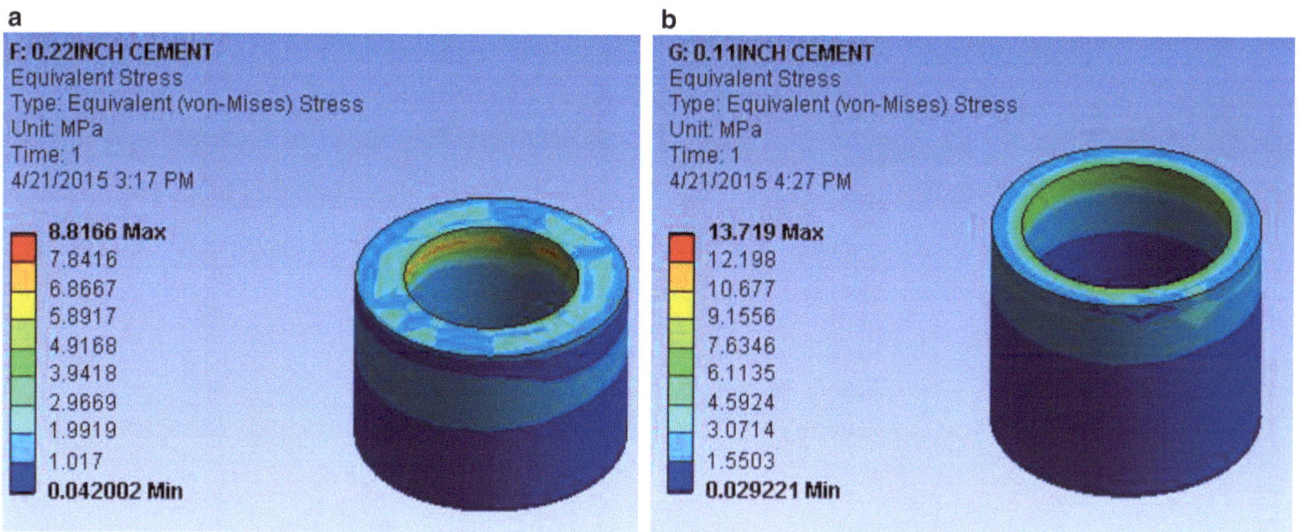

Fig. 1.7 (**a**) Von-Mises stress contour plot of 0.22 in. cement; (**b**) Von-Mises stress contour plot of 0.11 in. cement

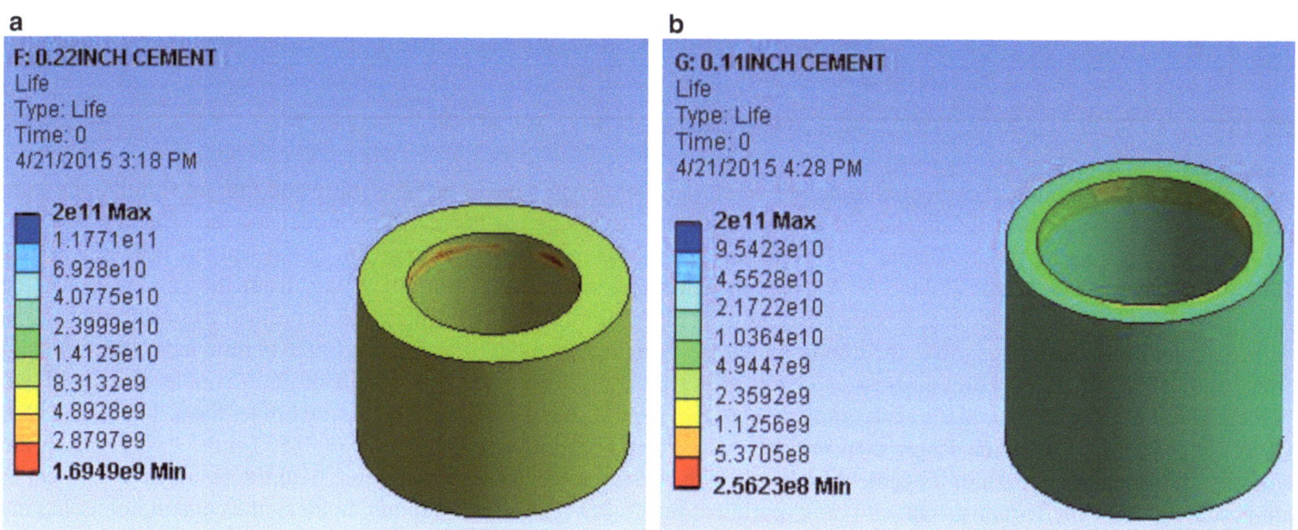

Fig. 1.8 (**a**) Fatigue life contour plot of 0.22 in. cement; (**b**) fatigue life contour plot of 0.11 in. cement

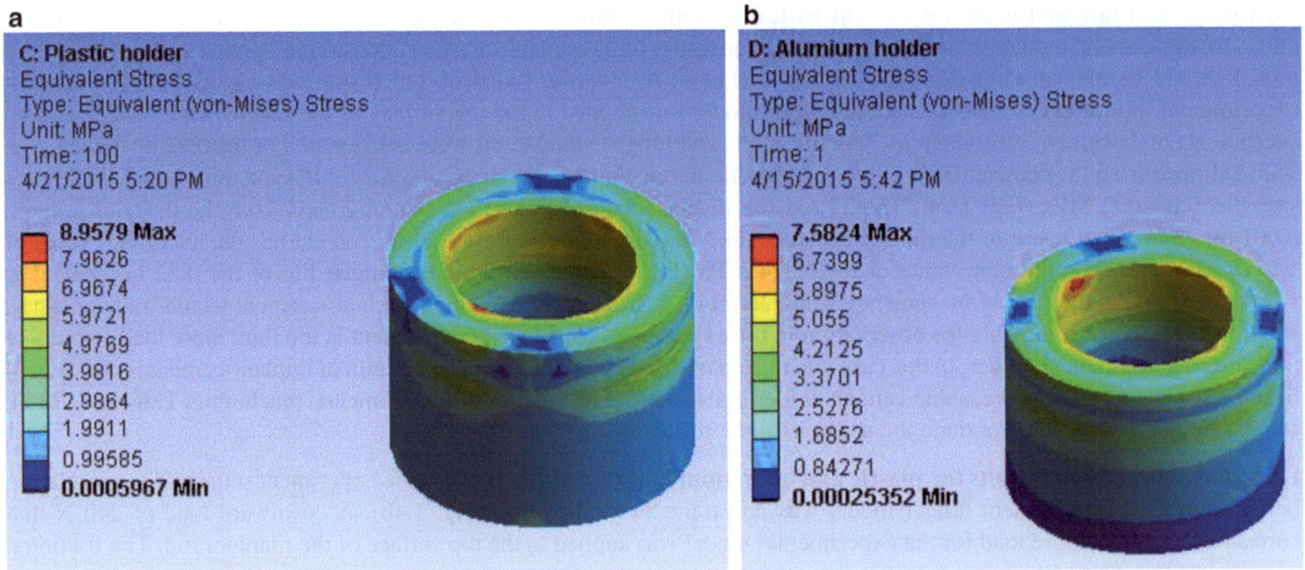

Fig. 1.9 (**a**) Von-Mises stress contour plot of cement in plastic holder; (**b**) Von-Mises stress contour plot of cement in aluminum holder

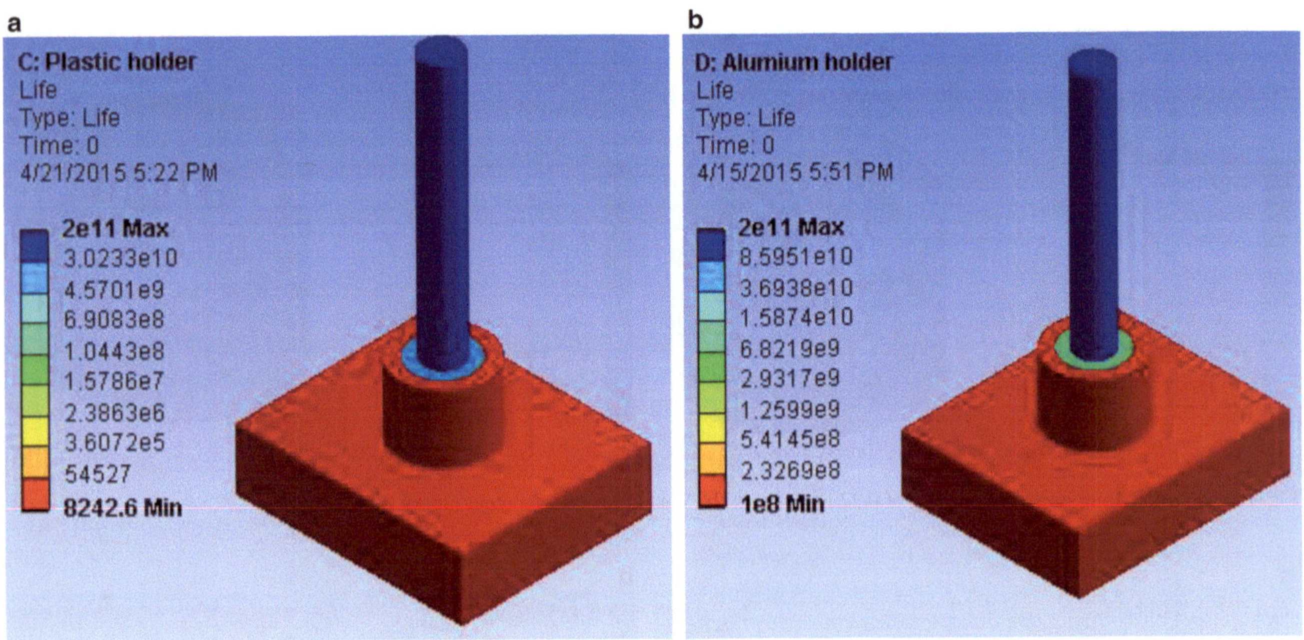

Fig. 1.10 (**a**) Fatigue life contour plot of the Ti-cement-plastic holder union; (**b**) fatigue life contour plot of the Ti-cement-aluminum holder union

coefficients of Ti-cement and cement-aluminum holder were input as 7 and 14 and the static and fatigue simulations were performed. FEA results show the displacement contour plot of the Ti-cement-holder union model and the cement only under the compressing load of 240 N. Figure 1.9a, b display the Von-Mises stress contour plot of cement for both plastic and aluminum holders. For plastic holder, Ti rod has displacement of 0.030457 mm at Z-axis from the cement under the minimum stress of 0.5967 kPa and maximum stress of 8.9579 MPa (Fig. 1.9a). Under the same load as plastic model, the deformation and the stress was lower for aluminum holder. The displacement of Ti rod was 0.01649 mm under the minimum stress of 0.25352 kPa and maximum stress of 7.5824 MPa (Fig. 1.9b). Comparing the results of plastic holder and aluminum holder, Ti rod of the plastic holder transfer more load to the cement so the cement of the plastic holder has a displacement of 0.013529 mm, which is much larger than that of the cement with aluminum holder 0.0088523 mm. According to the analysis, the fracture strength of Ti-cement interface with plastic holder has a lower value than the result of Ti-cement in aluminum holder, which is consistent with the experiment result. For plastic or aluminum holders, the cement holder has the minimum design life of the whole setup (Fig. 1.10). Plastic holder has only 8242.6 cycles (Fig. 1.10a), which suggested the plastic holder is the most subject to damage. The cement in the plastic holder has the minimum design life of the 1.0443e8 cycles. For the aluminum holder, the cement holder also has the minimum design life of the whole setup, 1e8 (Fig. 1.10b), much more than the plastic holder. Moreover, the cement of the aluminum holder has a longer fatigue life of 2.9317e9 cycles, around 20 times more than the cement of the plastic holder. Based on the result, plastic holder is not suitable for fatigue test. To avoid the cement holder damaged before the fracture of the Ti-cement interface, aluminum is the better option as the cement holder.

1.4 Conclusions

This study evaluated the effect of cement thickness and cement holder material on the fracture of Ti-cement interface by experiment and FEA. This result shows that the cement thickness of 0.22 in. has more fracture strength and fatigue life than the cement thickness of 0.11 in.. Either plastic or aluminum holder can be used for static test. However, the fracture strength result got from the model with plastic holder was much less than the result with aluminum holder. The plastic holder is the most subject to damage during the fatigue test. Thus plastic holder is not better that aluminum holder for fatigue test.

Acknowledgements This research was made possible by grant 5P20GM103447 from the US National Institutes of Health and an on-campus faculty grant program from the University of Central Oklahoma Office of Research and Grants.

References

1. Davies, J., Harris, W.: Strength of cement-metal interfaces in fatigue: comparison of smooth, porous and precoated specimens. Clin. Mater. **12** (2), 121–126 (1993)
2. Fisher, D.A., Tsang, A.C., Paydar, N., Milionis, S., Turner, C.H.: Cement-mantle thickness affects cement strains in total hip replacement. J. Biomech. **30**(11/12), 1173–1177 (1997)
3. Biomet, I.: Cobalt™ HV Bone Cement. http://www.biomet.com/orthopedics/productDetail.cfm?category=7&product=250 (2010)
4. Stratasys Inc., "Dimension-Elite Product Review", www.stratasys.com/3d-printers/design-series/dimension-elite, (July 12, 2016)
5. Test Resources Inc. "Model 800LE4 overview" http://www.testresources.com/accessories/AccDetail.aspx?pid=219&fam=21&pritype=Exten&stype=Eoption, (July 12, 2016)
6. Ries, M.D., et al.: Intramedullary pressure and pulmonary function during total knee arthroplasty. Clin. Orthop. Relat. Res. **356**, 154–160 (1998)
7. ANSYS Inc. www.ansys.com (2016)
8. Huiskes, R.: The various stress patterns of press-fit, ingrown and cemented femoral stems. Clin. Orthop. **261**, 27–38 (1990)
9. Li, C., Mason, J., Yakimicki, D.: Thermal characterization of PMMA-based bone cement curing. J. Mater. Sci. Mater. Med. **15**(1), 85–89 (2004)
10. Nuño, N., Avanzolini, G.: Residual stresses at the stem–cement interface of an idealized cemented hip stem. J. Biomech. **35**(6), 849–852 (2002)

Chapter 2
Mechanical Behavior and Aluminization of Cu21Zn6Al Alloys

H.H. Li and Q.Z. Li

Abstract Copper-21%zinc-6%aluminum (Cu21Zn6Al) alloys were produced through induction casting and processed using various heat treatment methods to investigate their microstructure and mechanical behavior. Some samples were homogenized for different durations, heated for some duration, and quenched by four types of routes. The other samples were heated for different durations and quenched by a two-step quenching. One sample was aluminized to study the aluminization process of the alloy. The processed alloys were studied using optical microscope and scanning electron microscope to explore the microstructure. A lamellar microstructure was observed. The Vickers hardness data were obtained using a microhardness tester. The results show that the grain size increases with the increase of the homogenization duration; the sample quenched to ice directly experienced the fastest cooling rate and thus has the smallest grain size compared with the other three types of quenching routes; the sample experienced longer homogenization duration has lower hardness value and the lowering rate of hardness value levels off with the increase of homogenization duration; the increase of heating temperature from 900 to 950 °C resulted in the decrease of hardness value; and the aluminization process realized an aluminized protection layer on the sample surface.

Keywords Copper-zinc-aluminum alloy • Homogenization • Quenching • Microhardness • Aluminization

2.1 Introduction

Cu alloys have a wide range of practical applications such as architecture, seawater, tubes, pipes and fittings because they have high electric conductivity and good machinability, and especially good corrosion resistance and capability to stand the harsh seawater environment [1]. It is well known that corrosion is one of the most commonly seen failure mode of metallic components and the direct loss of metallic corrosion in the United States of America is about $276 billion every year (i.e. almost 3.1 % of the annual gross domestic product) [2]. The addition of Al can contribute to the improvement of corrosion resistance. There are also many ways in the industries to apply a corrosion resistant coating on the metallic surface such as carburization, nitronization, chromization, aluminization, and siliconization. Aluminization was widely utilized to form a coating on a range of different iron and steel-based alloys, nickel-based alloys, and copper-based alloys [3–10]. In this study, Zn and Al were added to Cu to produce Cu21Zn6Al alloy, various heat treatment processes were performed to explore the effect of heat treatment on microhardness of the Cu21Zn6Al alloy, and aluminization was performed to study the coating on the Cu21Zn6Al alloy.

2.2 Experimental Methods

The Cu21Zn6Al samples were prepared using 99.9 % pure Cu, 99.99 % pure Zn, and 99.99 % pure Al in an induction furnace. Four groups of heat treatment were performed on the samples as listed in Table 2.1. For Group No. 1, the samples were homogenized at 500 °C for 2, 4, and 6 h respectively and then heated at 900 °C for 15 min followed by quenching to ice.

H.H. Li
University of Nevada-Reno, Reno, NV 89557, USA

Q.Z. Li (✉)
University of Nevada-Reno, Reno, NV 89557, USA

Washington State University, Pullman, WA 99164, USA
e-mail: qizhen.li@wsu.edu

Y. Zhu, A.T. Zehnder (eds.), *Experimental and Applied Mechanics, Volume 4*,
Conference Proceedings of the Society for Experimental Mechanics Series, DOI 10.1007/978-3-319-42028-8_2

Table 2.1 Four groups of heat treatment processes performed on the Cu21Zn6Al samples

Group no.	Homogenization temperature (°C)	Homogenization duration	Heating temperature (°C)	Heating duration (min)	Cooling
1	500	2, 4, 6 h	900	15	Ice
2	500	2 h	900	15	Route 1: quenching to ice
					Route 2: quenching to the boiling water for 30 min and then to ice
					Route 3: quenching to water at room temperature
					Route 4: quenching to the boiling water for 30 min and then to water at room temperature
3	–	–	900	5, 10, 15, 20, 25, and 30	Two-step quenching: to the boiling water for 30 min and then to water at room temperature
4	–	–	950	5, 10, 15, 20, 25, and 30	Two-step quenching: to the boiling water for 30 min and then to water at room temperature

For Group No. 2, the samples were homogenized at 500 °C for 2 h and then heated at 900 °C for 15 min followed by quenching through four routes. Route 1 is quenching to ice, route 2 is quenching to the boiling water for 30 min and then to ice, route 3 is quenching to water at room temperature, and route 4 is quenching to the boiling water for 30 min and then to water at room temperature. For Group No. 3, the samples were heated at 900 °C for 5, 10, 15, 20, 25, and 30 min respectively and then experienced two-step quenching. For Group No. 4, the samples were heated at 950 °C for 5, 10, 15, 20, 25, and 30 min respectively and then experienced two-step quenching. During the two-step quenching, the samples were first quenched to the boiling water for 30 min and then to water at room temperature.

The samples were ground, polished, and etched using the NH_4OH and H_2O_2 solution for microstructure observation. Visual light microscope (VLM) and scanning electron microscope (SEM) were utilized for the microstructure characterization. The energy dispersive X-ray spectroscopy (EDX) was used for the elemental analysis. Microhardness data were collected for the samples experienced quenching.

A sample was aluminized inside a package containing 95 wt% Al_2O_3, 1.5 wt% NH_4Cl, and 3.5 wt% Al powders at 900 °C for 5 h. After aluminization, the sample was examined under SEM to explore its microstructure and composition.

2.3 Results and Discussion

Figure 2.1a shows the SEM image of a Cu21Zn6Al sample and a point was scanned to collect the EDX data as shown in Fig. 2.1b. The data indicate the existence of Cu, Zn, and Al in the sample. Figure 2.2 reports the optical micrographs of the as-cast Cu-21Zn-6Al alloy. The SEM image in Fig. 2.1a and the optical micrograph in Fig. 2.2b show the lamellar structure in the alloy.

Figure 2.3 shows the optical micrographs of the samples in the second group of samples that were homogenized at 500 °C for 2 h and then heated at 900 °C for 15 min followed by quenching through four routes. The grain size after Route 1 (i.e. quenching directly to ice) is the smallest since this quenching route has the fastest cooling rate, while that after Route 4 (i.e. quenching to the boiling water for 30 min and then to water at room temperature) is the largest since this quenching route has the slowest cooling rate. The grain size after Route 1 is smaller than that after Route 2 (i.e. quenching to the boiling water for 30 min and then to ice), and the grain size after Route 3 (i.e. quenching to water at room temperature) is smaller than that after Route 4. The addition of quenching to the boiling water for 30 min resulted in the increase of grain size. Figures 2.3a and 2.4 present the optical micrographs of the first group of samples that were homogenized at 500 °C for 2, 4, and 6 h respectively and then heated at 900 °C for 15 min followed by quenching to ice. The images indicate that the samples experienced the longer homogenization duration have larger grain size.

Microhardness data were collected and reported in Figs. 2.5 and 2.6. The data in Fig. 2.5a indicate that the sample experienced longer homogenization time has lower hardness value. The lowering rate of hardness value levels off with the increase of homogenization duration. The hardness value after Route 1 is the largest, while that after Route 2 is the smallest. The hardness value after Route 1 is larger than that after Route 2, and the hardness value after Route 3 is larger than that after Route 4. The addition of quenching to the boiling water for 30 min resulted in the decrease of hardness value. Figure 2.6 shows that the increase of heating temperature from 900 to 950 °C resulted in the decrease of hardness value. The overall trend is that samples stayed for longer duration at the heating temperature have lower hardness values.

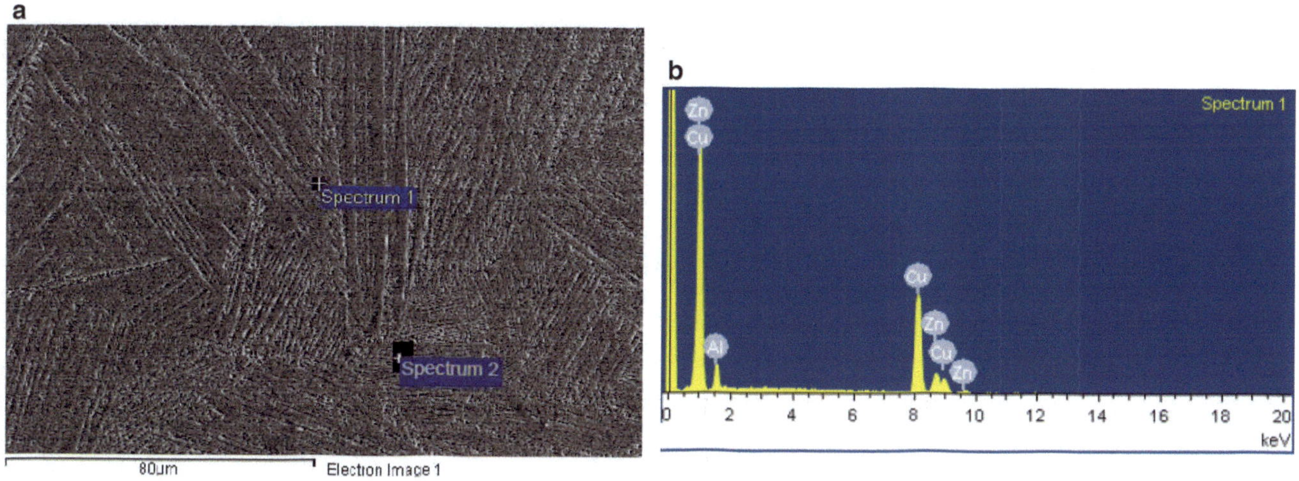

Fig. 2.1 (**a**) SEM image and (**b**) EDX analysis of Cu21Zn6Al alloy

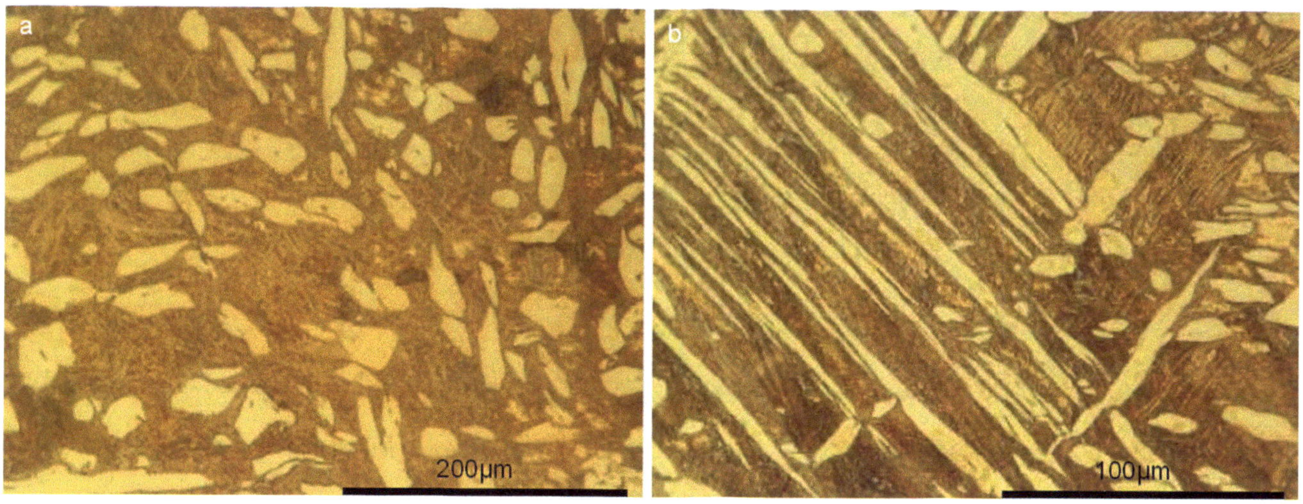

Fig. 2.2 Optical micrographs of Cu21Zn6Al alloy at (**a**) low magnification and (**b**) high magnification

The Cu21Zn6Al sample after the aluminization processing for 5 h was studied using SEM-EDX. Figure 2.7 reports the SEM-EDX result with the sample surface on the left edge. A horizontal line scan to the inside of the sample shows the composition variation of Cu and Al. There was an obvious drop in Al when the scanning was across the boundary and moved to the right of the boundary, which is related to the aluminization processing. This data indicate that an aluminized protection layer was formed on the sample surface.

2.4 Conclusion

This work produced Cu21Zn6Al alloys through induction casting and processed them through four groups of heat treatment methods. Additionally, one sample was aluminized to study the aluminization process of the alloy. The samples were characterized microstructurally and mechanically using VLM, SEM, EDX, and microhardness tester. The microstructure characterization results show that there is lamellar microstructure in the as-cast state; the grain size increases with the increase of the homogenization duration; and the grain size is the smallest for the sample quenched to ice and it is the largest for the sample quenched to the boiling water for 30 min and then to water at room temperature. The microhardness testing results show that the sample experienced longer homogenization duration has lower hardness value; the hardness decrease

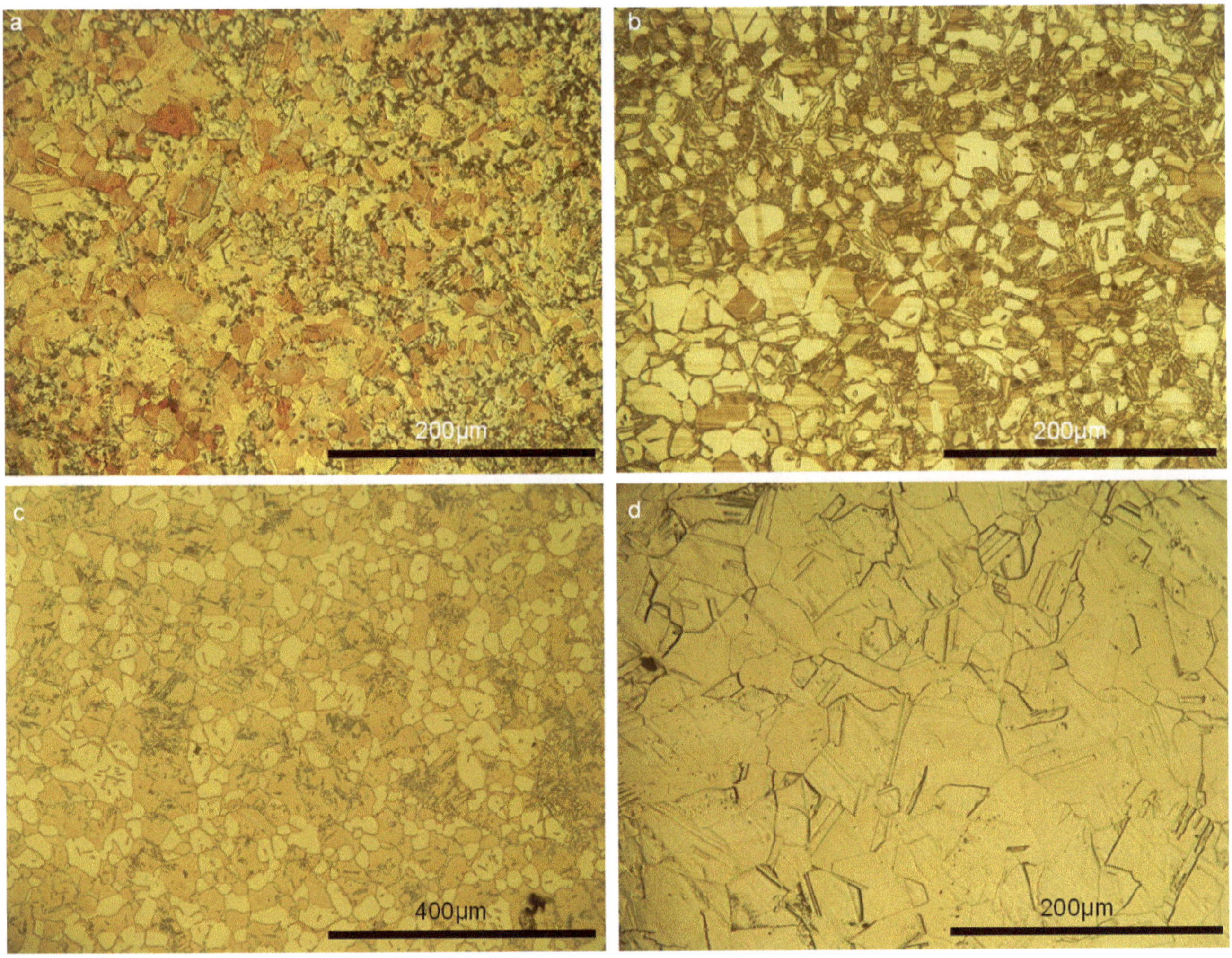

Fig. 2.3 Optical micrographs of Cu21Zn6Al alloy homogenized at 500 °C for 2 h and then heated at 900 °C for 15 min followed by quenching through four routes. (a) Route 1: quenching to ice, (b) route 2: quenching to the boiling water for 30 min and then to ice, (c) route 3: quenching to water at room temperature, and (d) route 4: quenching to the boiling water for 30 min and then to water at room temperature

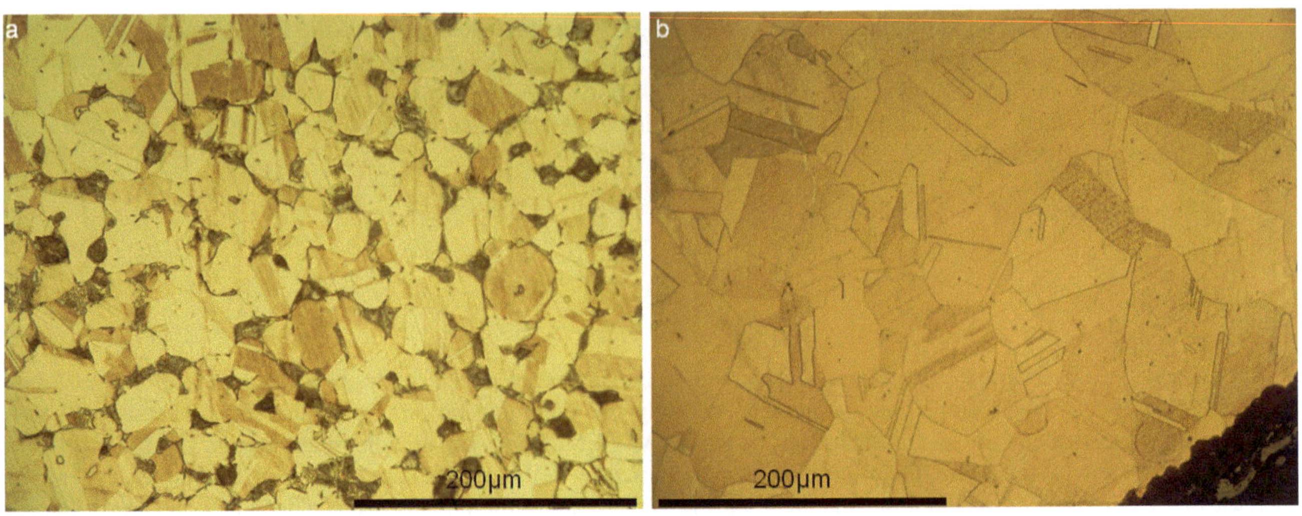

Fig. 2.4 Optical micrographs of Cu21Zn6Al alloy homogenized at 500 °C for (a) 4 h and (b) 6 h and then heated at 900 °C for 15 min followed by quenching to ice

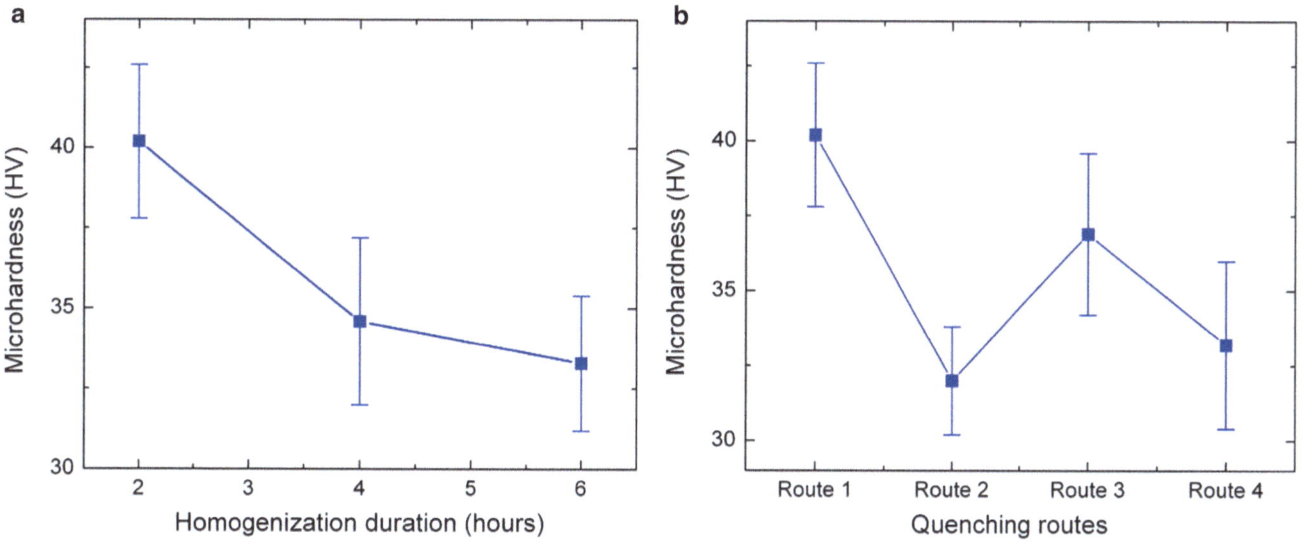

Fig. 2.5 (**a**) Microhardness of Cu21Zn6Al alloy homogenized at 500 °C for 2, 4, and 6 h respectively and then heated at 900 °C for 15 min followed by quenching to ice. (**b**) Microhardness of Cu21Zn6Al alloy homogenized at 500 °C for 2 h and then heated at 900 °C for 15 min followed by quenching through four routes

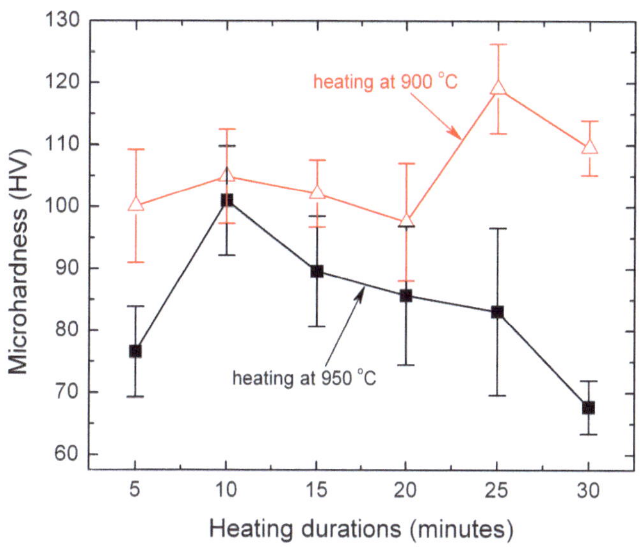

Fig. 2.6 Microhardness of Cu21Zn6Al alloy heated at different high temperatures (900 and 950 °C) for 5, 10, 15, 20, 25, and 30 min respectively and then experienced two-step quenching

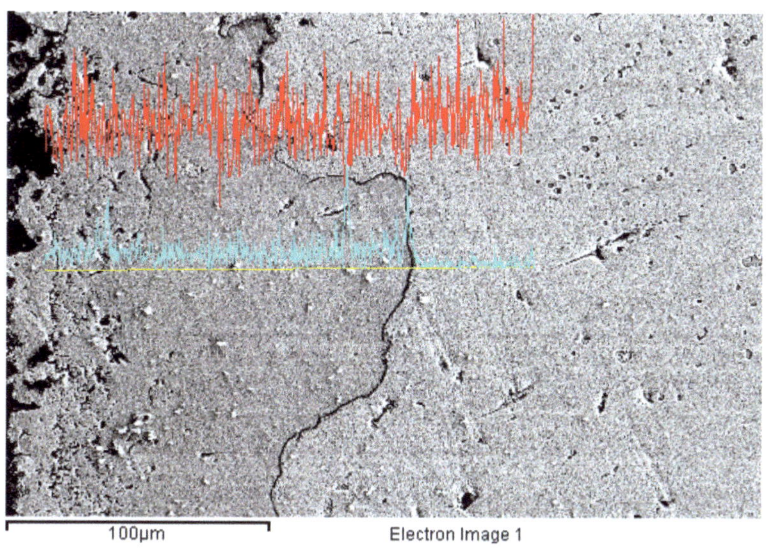

Fig. 2.7 SEM image and line scanning EDX data of the Cu21Zn6Al alloy after the aluminization processing for 5 h

rate levels off with the increase of homogenization duration; and the increase of heating temperature from 900 to 950 °C resulted in the decrease of hardness value when the other processing parameters are fixed. The microstructure and composition analysis also indicates that the aluminization process formed an aluminized protection layer on the sample surface.

References

1. Davis, J.R.: ASM Specialty Handbook: Copper and Copper Alloys. ASM International, Materials Park (2001)
2. Koch, G.H., Brongers, M.P.H., Thompson, N.G., Paul Virmani, Y., Payer, J.H.: Corrosion costs and preventive strategies in the United States. NACE International (2002)
3. Kircher, T.A., McMordie, B.G., Richards, K.: Use of experimental designs to evaluate formation of aluminide and platinum aluminide coatings. Surf. Coat. Technol. **108–109**, 24–29 (1998)
4. Rairden, J.R., Jackson, M.R.: A cobalt surface pretreatment for Ni-Cr-type alloys to attain pore-free aluminized coatings. Thin Solid Films **64**, 291–297 (1979)
5. Su, C., Lee, J.W., Wang, C.S., Chao, C.G., Liu, T.F.: The effect of hot-dipped aluminum coatings on Fe-8Al-30Mn-0.8C alloy. Surf. Coat. Technol. **202**, 1847–1852 (2008)
6. Tsai, W., Huang, K.: Microstructural aspect and oxidation resistance of an aluminide coating on 310 stainless steel. Thin Solid Films **366**, 164–168 (2000)
7. Sanchez, L., Bolivar, F.J., Hierro, M.P., Perez, F.J.: Effect of Ce and La additions in low temperature aluminization process by CVD-FBR on 12% Cr ferritic/martensitic steel and behaviour in steam oxidation. Corros. Sci. **50**, 2318–2326 (2008)
8. Nciri, B., Vandenbulcke, L.: Theoretical and experimental study of the aluminization of iron and steel in the pack and in the gas phase. Thin Solid Films **139**, 311–324 (1986)
9. Gupta, B.K., Seigle, L.L.: The effect on the kinetics of pack aluminization of varying the activator. Thin Solid Films **73**, 365–371 (1980)
10. Chiang, K.T., Kallenborn, K.J., Yuen, J.L.: Aluminization of copper for oxidation protection. Surf. Coat. Technol. **52**, 135–139 (1991)

Chapter 3
Numerical Analysis of Stress/Strain Fluctuations in Coiled Tubing During Deepwater Deployment

Joanne Ishak and Steven M. Tipton

Abstract Conventional coiled tubing experiences a set of bending and straightening cycles during deployment into and out of a well bore. The severe bending strains that must be imposed on the tubing can be up to 3 % and are therefore well above the yield strength of the material. This leads to fatigue failure at extremely short lives.

However, during deep water well interventions coiled tubing is also subjected to high cycle fatigue. After exiting the sheave, the tubing is deployed through open water with a clump weight and suspended vertically at water depths up to 3800 m. This happens while fluid is being pumped at pressures that can reach 68 MPa. The ocean waves impose pitch and roll on the vessel. This causes wrapping and unwrapping motion of the tubing on and off the sheave, which induces stress/strain fluctuations at the tangent point and can lead to high cycle fatigue damage accumulation.

To understand the stress state at the critical location, detailed finite element analyses were conducted using sophisticated incremental plasticity and contact elements to quantify the influence of the angular displacement magnitude, the axial force, and the internal pressure on the stress range at the critical location. Four cases were investigated: low pressure/medium force, high pressure/medium force, low pressure/high force, and high pressure/high force. When comparing the maximum principal strain ranges at small pitching angle ($\leq 2°$), it has been found that at a fixed internal pressure, the same strain strange is depicted for both axial forces. At higher pitching angles, however, the strain range increases with increasing axial force. In addition, axial forces tend to have a greater effect on the maximum principal strain range at a smaller pressure.

Keywords Finite element analysis • High cycle fatigue • Plastic regime • Pitch angle • Coiled tubing

3.1 Introduction

Coiled Tubing is widely used in the oil field industry due to considerable cost and time savings over more conventional drill pipe methods [1–3]. The continuous string of steel tubing is initially wound on a service spool. During operation, the tubing is unwound (straightened) from the spool, bent again as it passes over a guide arch into an injector which lowers the tubing into the well, or in some cases, pushes it into the well. After completing its downhole task, the tubing is pulled out of the bore and bent back over the guide arch, where it again straightens before wrapping onto the service spool. This routine operation imposes a discrete set of bending and straightening cycles. Bending strains associated with each bending-straightening cycle are on the order of 2–3 % [4, 5], and are therefore well beyond the elastic limit of the material, forcing it into the plastic regime. These events occur while fluids are being pumped at high internal pressures (up to 68 MPa). Consequently, during constant pressure operations the tubing tends to grow in diameter. In the lab, the tubing diameter can grow as much as 30 %.

Combined together, these loadings can cause the tubing to fail at lives fewer than 100 cycles, and therefore operate in the ultra-low cycle fatigue regime. Because of this, intensive research [4, 6–10] has been conducted to investigate the fatigue behavior of conventional coiled tubing.

Low cycle fatigue is not, however, the only fatigue regime experienced by coiled tubing. When the tubing is used in deep-water well interventions, it undergoes high cycle fatigue damage accumulation induced by ocean waves. The operational procedure consists of suspending the tubing over a sheave either through a moon pool or over the side of the vessel.

In addition to high internal pressures and large bending strains imposed on the tubing, a clump weight is attached to the tubing during its deployment through open waters to suspend it vertically at depths up to 3800 m. As the tubing hangs

J. Ishak (✉)
The University of Tulsa, Stephenson Hall 2165, Tulsa, OK 74104, USA
e-mail: joanne-ishak@utulsa.edu

S.M. Tipton Ph.D., P.E., FASME, FASM
The University of Tulsa, Stephenson Hall 2060, Tulsa, OK 74104, USA

Y. Zhu, A.T. Zehnder (eds.), *Experimental and Applied Mechanics, Volume 4*,
Conference Proceedings of the Society for Experimental Mechanics Series, DOI 10.1007/978-3-319-42028-8_3

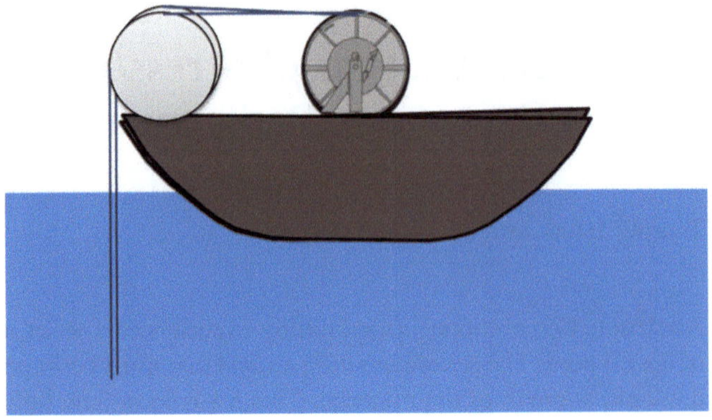

Fig. 3.1 Vessel motion

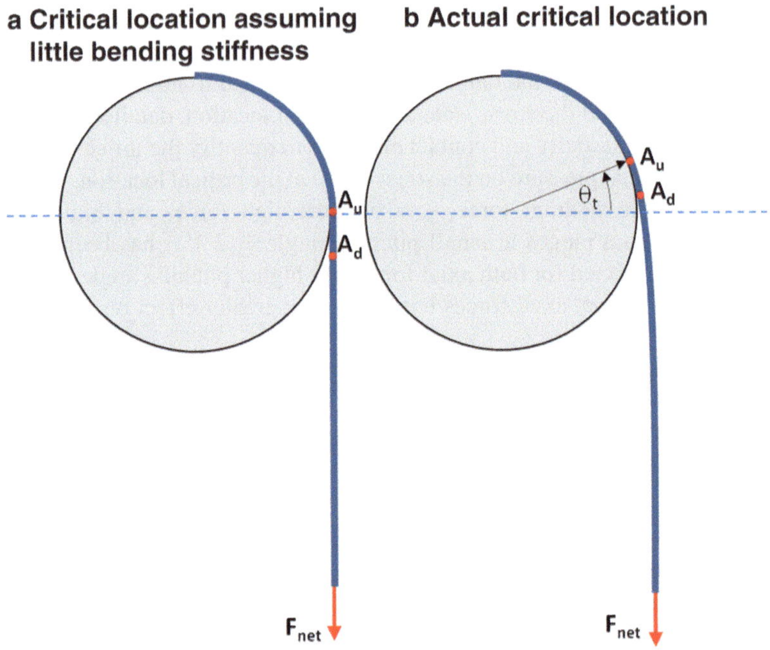

Fig. 3.2 Hot spot motion

through open water, all 6° of vessel motion can impose flexural strains on the tubing at the critical location. However, it is the pitch or roll of the vessel (depending on how the axes of the spool and sheave are oriented relative to the vessel) that causes similar rotation of the spool, leading to small amounts of wrapping and unwrapping at the critical location; which is the point at which the pipe becomes tangent to the sheave. This is illustrated in Figs. 3.1 and 3.2. Figure 3.1 shows a situation where the axes of the spool and sheave are aligned with the pitch axis of the vessel. When deployed over the side, the roll axis is aligned with the axes of the spool and sheave.

The critical location on the tubing is defined as the point where the tubing exits the sheave when the sheave is rotated into its full upward position by the angular motion of the vessel. This location is illustrated by point A in Fig. 3.2 in two different positions of the vessel: when the vessel is oriented in its full upward and downward positions. If the tubing had little bending stiffness, like limp rope, the critical location A_u would be defined by the horizontal axis through the sheave, as shown in Fig. 3.2a. However, the stiffness of the tubing causes it to exit the sheave at point A_u at some position defined by θ_t as shown in Fig. 3.2b. When the vessel rocks by some amount $\Delta\theta$ to it full downward position, point A moves to location A_d. The distance between A_u and A_d is approximately $R\Delta\theta$, where R is the radius of curvature of the tubing centroidal axis when wrapped over the sheave.

In Fig. 3.2a, the bending strain in point A would fluctuate from that associated with the full curvature of the spool at A_u, to zero at A_d. This would lead to the almost immediate failure of the tubing (in a few hundred cycles or less). However, ample anecdotal data exist from case studies in the field where coiled tubing has been successfully utilized for such applications for periods of as long as 5 days. With average wave periods of 5 s, this represents 86,400 cycles over 5 days without failure.

Therefore a smaller strain fluctuation is occurring due to the stiffness of the tubing. Figure 3.2b illustrates this. In the UP position, A_u, the strain at point A is equal to the strain caused by the full curvature over the sheave (=c/R, where c is half the outer diameter of the tubing). As point A moves to position A_d, the curvature of the tubing decreases, but does not go to zero (fully straight) as shown in Fig. 3.2a.

The purpose of this study is to define the flexural strain at point A as it moves between positions A_u and A_d, and how this fluctuation is affected by the axial force, F_{net}, acting on the tubing and the internal pressure during the oscillation. The strain fluctuation can then be used for assessment of high cycle fatigue damage at this location. For this purpose, finite element analyses were conducted using the commercially available FEA software, Abaqus.

It should be noted that the vertical heave of the vessel, specifically at the location just above the critical location, also causes fluctuating axial stress in the tubing due to inertial loading. However, only the flexural component of strain (stress) fluctuation is being considered in this study. Analysis has shown that the inertial component of stress fluctuation is small compared to the flexural component, although the two could easily be combined in practice.

3.2 Finite Element Analysis

Finite element analyses were conducted to model the oscillating sheave rotation, assumed to be equivalent to the vessel angular displacement. This is done in order to quantify the flexural strain at the critical location (point A in Fig. 3.2b) depending on multiple variables. Several runs were performed by varying each of the following parameters: Pitch angle, $\Delta\theta$, tubing internal pressure, P, and tensile load due to clump weight and pipe's own weight. The input variables are shown in Table 3.1 below. It should be noted that F corresponds to a buoyed clump weight of 227 kg and depths of 3048 and 3810 m for the tubing size analyzed in this study.

3.3 Input Data

The geometries as well as the dimensions of the sheave and pipe are depicted in Figs. 3.3 and 3.4 below. Symmetry boundary conditions and some other simplifications in the sketches are applied to both structures to reduce computation time. Adequate boundary conditions are applied to both sketches to account for symmetry and other simplifications.

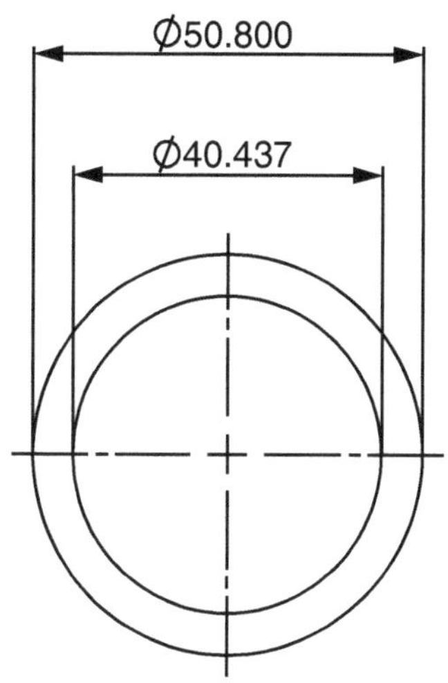

Fig. 3.3 Coiled tubing cross section in mm

Table 3.1 Input data

$\Delta\theta$ (°)	2	4	6
F_{net} (N)	115,653	191,273	–
Internal pressure (MPa)	1.38	68.95	–

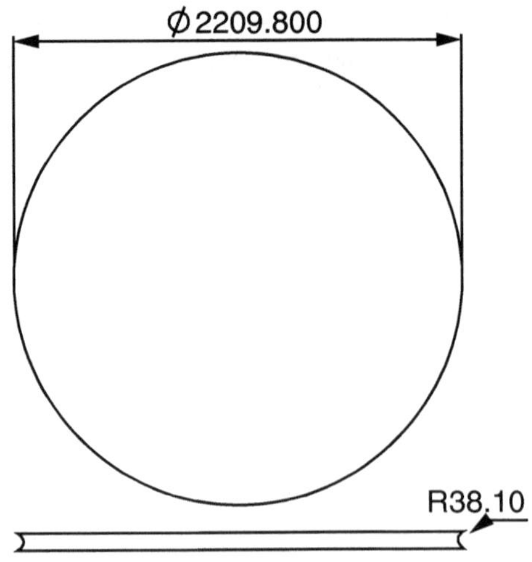

Fig. 3.4 Sheave dimensions in mm

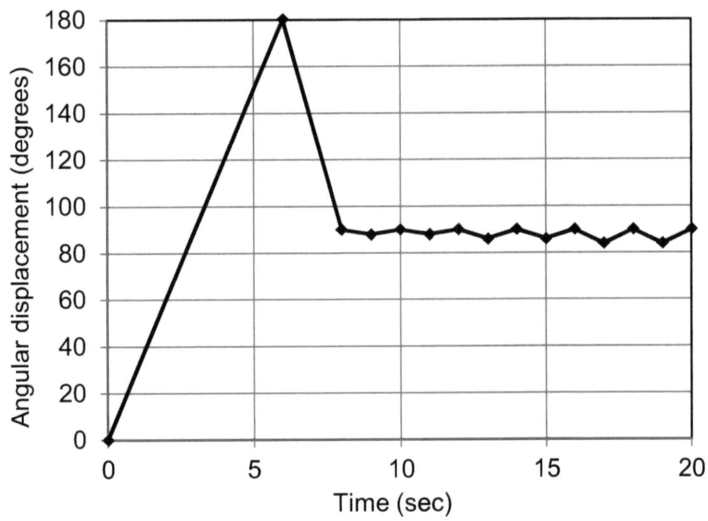

Fig. 3.5 Sheave angular displacement

The sheave is assumed to be a rigid body; hence no material properties are assigned to it. The pipe, however, is an alloy steel deformable 3D solid body having a Young's modulus of elasticity of 186,158 MPa (27,000,000 psi) and a Poisson's ratio of 0.3.

The plasticity model used is the bi-linear kinematic hardening model having a yield strength of 621 MPa (90,000 psi). The non-linear combined (isotropic/kinematic) hardening model is usually the most adequate model for modeling the behavior of metals subjected to cyclic loading. However, since the isotropic component that accounts for the cyclic hardening/softening behavior is not depicted in the case of high cycle fatigue regime, the combined model was not required in this study.

The sheave orientation history used to model the process is shown in Fig. 3.5. The tubing is wrapped around a 180° angle then a portion of the tubing is straightened from the sheave while keeping the remaining part wrapped around approximately a 90° angle. This is done to account for the bending-straightening cycle that occurs when the tubing is initially wound on the spool then straightened and extended on the sheave before being deployed. The tubing is then subjected to two 2° cycles, then two 4° cycles, and lastly two 6° cycles. The cyclic motion history is summarized as 0°–180°–90°–88°–90°–88°–90°–86°–90°–86°–90°–84°–90°–84°–90°.

3.4 Results and Discussion

The assembly and the maximum principal strain gradient during the first cycle (0°-180°) are shown in Fig. 3.6. Maximum strains are observed at the top part of the tubing due to pure bending. The axial force and internal pressure are applied in subsequent steps. The maximum principal strain histories for each pressure/axial force combination are illustrated in Fig. 3.7.

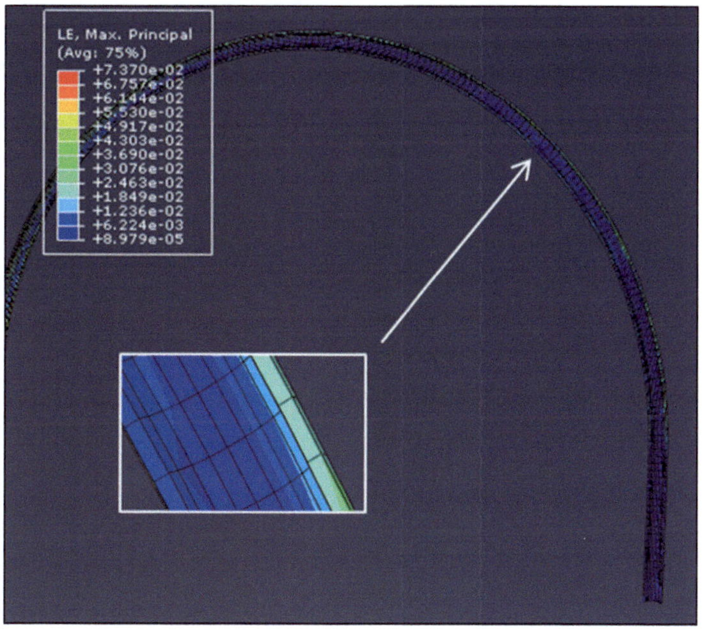

Fig. 3.6 Strain gradient

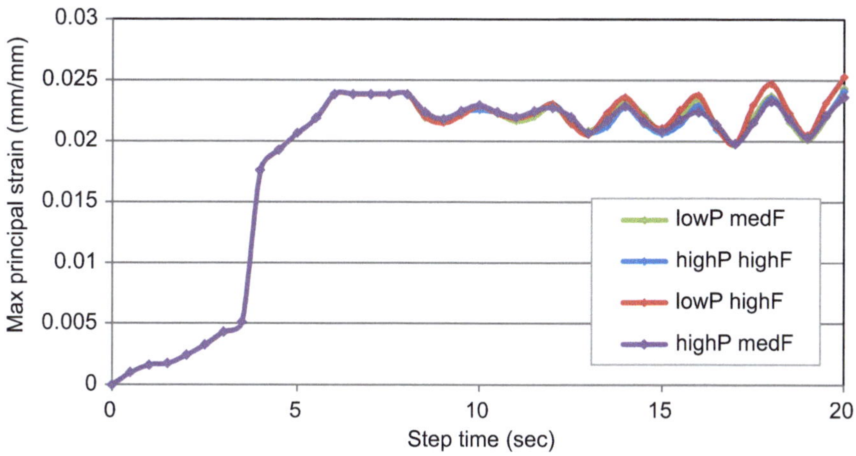

Fig. 3.7 Strain histories at the hot spot for each pressure-force combination

The strains depicted during the first two cycles (0°–180°–90°) are the same for all four combinations since external P and F loadings are not yet applied. Cyclic angular displacements are then applied to model the 2°, 4°, and 6° pitch angles. During these fluctuating displacements, the hot spot exits the sheave and then goes back to its initial location, where it becomes tangent to the sheave. The hot spot is located at the upper portion of the tubing and is identified as the last node where the strain is equal to c/ρ, where c is the distance from the neutral axis to the hot spot (half the tubing diameter) and ρ is the radius of curvature of the tubing wrapped on the sheave.

In order to investigate the effect of axial force on high cycle fatigue of coiled tubing, plots of $\Delta\varepsilon$ versus $\Delta\theta$ are shown in Figs. 3.8 and 3.9 for low and high pressures, respectively.

When comparing the curves at P = 1.38 MPa, it can be seen that at a small pitching angle (2°), the strain range is equal for the 26,000 lb and 43,000 lb forces. As the pitch angle increases, the strain range also increases. The percentage difference between the strain ranges is 0.84 %, 11.12 %, and 16.3 % for 2°, 4°, and 6° respectively.

This is also the case for P = 68.95 MPa. At 2° pitching angle, the strain range is the same at both axial forces. The percentage difference between $\Delta\varepsilon$ is 4.27 %, 15.54 %, and 8.62 % for 2°, 4°, and 6° respectively.

Comparing the strain range between low and high pressures indicates that larger strain ranges are observed at smaller internal pressure. The percentage difference of strain ranges between low and high pressure values are 29.17, 30.11, and 18.03 % for F = 115,653 N and 34.16, 25.75, and 25.65 % for F = 191,273 N.

Fig. 3.8 Strain range versus sheave angular displacement for P = 1.38 MPa (200 psi)

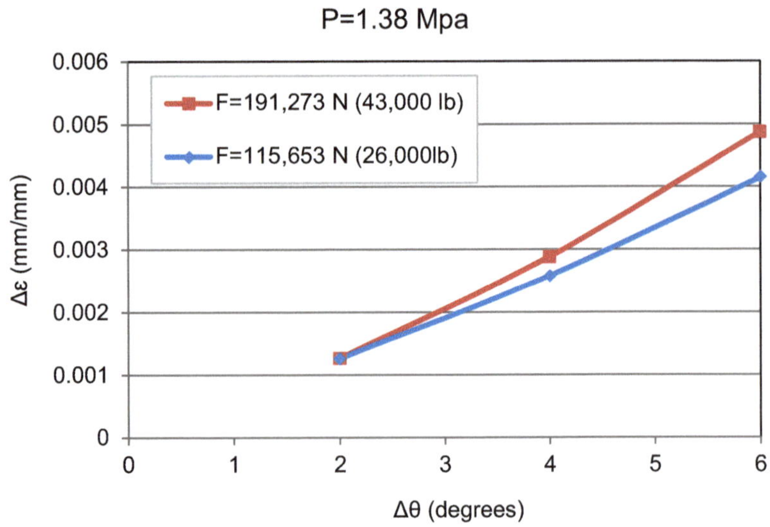

Fig. 3.9 Strain range versus sheave angular displacement for P = 68.95 MPa (10,000 psi)

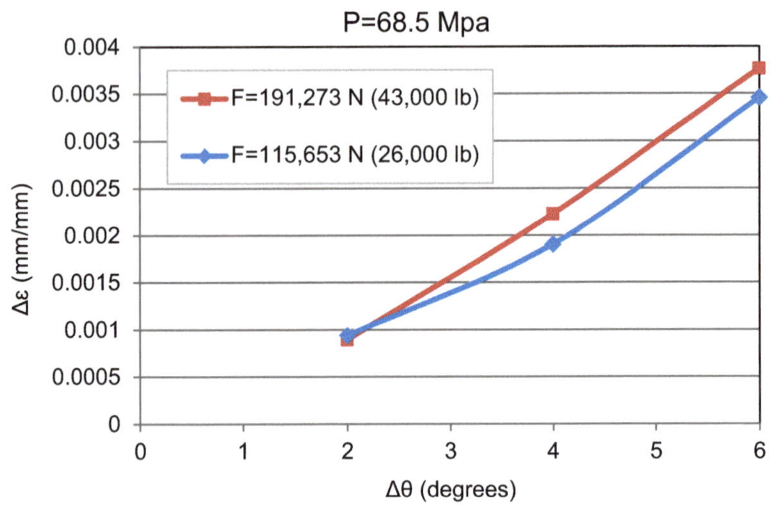

It is noteworthy that the high internal pressures, which leads to very short lives for coiled tubing cycled plastically, tends to reduce the fluctuation strain amplitude under these loading conditions, which would lead to higher estimated fatigue lives.

3.5 Conclusions

1. The magnitude of the axial force (tubing's weight + clump weight) does not affect the strain range at 2° pitch angle. Hence the strain range is the same at all water depths.
2. The strain range at the hot spot is larger for a higher axial force.
3. Larger pitch angles increase the strain range at the hot spot.
4. Increasing the internal pressure tends to reduce the strain range at the hot spot.

References

1. Hubb, A.: Crack propagation analysis of coiled tubing fatigue tests. Master's Thesis, The University of Tulsa, Tulsa (2014)
2. Kilambi, S.: Development of an algorithm for quantifying the dimensions of surface defects in coiled tubing from 3-dimensional laser scanning data. Master's Thesis, The University of Tulsa (2009)
3. Teel, M.E.: World Oil's Coiled Tubing Handbook (Reprinted). Gulf Publishing Company, Houston (1993)
4. Tipton, S.M.: Multiaxial plasticity and fatigue life prediction in coiled tubing. In: Mitchell, M.R., Landgraf, R.W. (eds.) Advances in Fatigue Lifetime Predictive Techniques: 3rd Volume, ASTM STP 1292, pp. 283–304. American Society for Testing and Materials, Philadelphia (1996)
5. Rolovic, R., Tipton, S.M.: Multiaxial cyclic ratcheting in coiled tubing—part I: theoretical modeling. J. Eng. Mater. Technol. **122**, 157–161 (2000)
6. Tipton, S., Newburn, D.: Plasticity and fatigue damage modeling of severely loaded tubing. In: Mitchell, M.R., Landgraf, R.W. (eds.) Advances in Fatigue Lifetime Predictive Techniques: 1st Volume, ASTM STP 1122, pp. 369–382. American Society for Testing and Materials, Philadelphia (1992)
7. Newburn, D.A., Tipton, S.M.: Influence of a dominant principal strain on low-cycle multiaxial fatigue. Paper presented at the fourth international conference on fatigue and fatigue thresholds, Honolulu, Hawaii, July 1990
8. Newburn, D.A.: Post yield cyclic strain response of pressurized tubes. Master's Thesis, the University of Tulsa, Tulsa (1990)
9. Tipton, S.M.: Coiled tubing deformation mechanics: elongation and diametral growth. Paper presented at the 1996 SPE/ICoTA 2nd North American Coiled Tubing Roundtable
10. Rolović, R.D., Tipton, S.M.: Elongation mechanisms in coiled tubing. Paper presented at the 1996 fourth international conference on coiled tubing operations, Gulf Publishing Company and World oil, Houston

Chapter 4
Determining SIFs Using DIC Considering Crack Closure and Blunting

G.L.G. Gonzáles, J.G. Diaz, J.A.O. González, J.T.P. Castro, and J.L.F. Freire

Abstract Three analytical-experimental hybrid approaches for determining the range of stress intensity factors (SIF) and the pseudo-SIFs of fatigue cracks on the presence of crack closure, crack tip plasticity and blunting are presented and evaluated. These approaches use the Digital Image Correlation (DIC) technique to measure cyclic-varying displacement fields near the crack tip. The first uses displacement data of points located on the cracked component surface near and along the crack faces, finding the parallel and orthogonal displacements of symmetrical points with relation to the crack faces to determine the crack opening displacements (COD), which are placed in the crack displacement field equations to obtain the SIF values. The second determines coefficients of generalized Westergaard displacement field functions by an over-deterministic nonlinear Least Square scheme that allows the accurate localization of the crack tip coordinates. The third computes the J-integral along an arbitrary but elastic contour path placed around the crack tip to determine the SIF, using experimentally determined displacement gradients, the calculated stresses and the calculated energy densities for points along the path. The three approaches are applied to the case of cracks propagating in disk-shaped compact-tension DC (T) specimens subjected to mode I cyclic loads, considering the non-linear effects mentioned above.

Keywords Digital image correlation • Fracture mechanics • Stress intensity factor • Crack opening displacement • J-integral

4.1 Introduction

This paper focuses on the evaluation of the stress intensity factor (SIF) of a fatigue cracked specimen using the digital image correlation (DIC) technique. The cases studied here involve non-linear behaviors induced by crack closure, by crack tip plasticity and blunting. Their effects on the displacement field led to the determination of the so-called pseudo-SIFs, since Linear Elastic Fracture Mechanics (LEFM) equations were employed in the SIF determinations.

In experimental mechanics, modern optical techniques such as digital image correlation (DIC) [1, 2] allow the calculation of LEFM parameters directly from the displacement fields experimentally measured around the crack tip. The idea of using the DIC displacements to determine SIFs was proposed by McNeill et al. [3] and since then, this method has been improved by many approaches that can be found in the literature, see e.g. [4–10].

Three common approaches are investigated here. In the first, the displacement components of symmetrical points located along the crack faces behind the crack tip are used to measure crack opening displacements (COD), and the SIF is calculated by placing the COD values in the displacement field equations obtained from the dominant near-crack-tip terms of the Westergaard solution for the stress state around the crack tip [11]. The second uses an over-deterministic Least Square Method (LSM) to analyze the full-field displacement data of the crack-tip region and to determine the coefficients of higher-order terms of the asymptotic expansion of the displacement components around the crack tip [12]. From the estimated coefficients, not only the SIF but also the crack-tip location can be extracted using a nonlinear least-squares approach [5]. In this work, an alternative searching method is adopted for localizing the crack tip coordinates based on the minimization of the error between the experimental displacements and the curve fitted displacements [8, 10]. This method converts the non linear problem on a sequence of linear least square problems that is easy to be implemented. In the third approach, a J-integral analysis is used to calculate the value of the SIF. The J-integral introduced by Rice [13] is computed directly from the measured displacement components along an arbitrary contour path placed around the crack tip. This methodology does not require that the crack tip to be accurately located, so it can be applied to the elastic region surrounding the crack tip.

G.L.G. Gonzáles • J.G. Diaz • J.A.O. González • J.T.P. Castro • J.L.F. Freire (✉)
Pontifical Catholic University of Rio de Janeiro, PUC-Rio, Rua Marquês de São Vicente 225, Gávea, Rio de Janeiro, RJ, Brazil
e-mail: jlfreire@puc-rio.br

Y. Zhu, A.T. Zehnder (eds.), *Experimental and Applied Mechanics, Volume 4*,
Conference Proceedings of the Society for Experimental Mechanics Series, DOI 10.1007/978-3-319-42028-8_4

These three approaches were implemented using the Matlab® software to process automatically the raw data from the DIC analysis.

The experimental part of the study consisted of determining DIC full-field displacements in the near-tip crack-region of a disk-shaped compact-tension specimen of an AISI 4340 steel tested under Mode I fatigue crack growth.

4.2 Calculating SIFs Using COD Measurements

Figure 4.1 shows an infinite cracked isotropic and linear-elastic plate, where x and y are the coordinates of an arbitrary point P in the coordinate system with origin at the crack tip; u and v are the displacement in x and y directions, respectively; r and θ are the polar coordinates.

The Westergaard solution [11] for the stress state near the crack tip can be used to find the in-plane displacement fields, u and v. By considering only its first terms, the displacement equations under mode I and mode II can be written as

$$u = \frac{K_I}{G}\sqrt{\frac{r}{2\pi}}\cos\frac{\theta}{2}\left[\frac{1-\mu}{1+\mu} + \sin^2\frac{\theta}{2}\right] + \frac{K_{II}}{G}\sqrt{\frac{r}{2\pi}}\sin\frac{\theta}{2}\left[\frac{2}{1+\mu} + \cos^2\frac{\theta}{2}\right] \tag{4.1}$$

$$v = \frac{K_I}{G}\sqrt{\frac{r}{2\pi}}\sin\frac{\theta}{2}\left[\frac{2}{1+\mu} - \cos^2\frac{\theta}{2}\right] + \frac{K_{II}}{G}\sqrt{\frac{r}{2\pi}}\cos\frac{\theta}{2}\left[\frac{1-\mu}{1+\mu} + \sin^2\frac{\theta}{2}\right] \tag{4.2}$$

where K_I and K_{II} are the mode I and mode II stress intensity factors respectively. The material elastic constant is the shear modulus $G = E/2(1 + \mu)$, where E is the Modulus of Elasticity and μ is the Poisson's ratio.

The model shown in Fig. 4.1 assumes a crack along the x-axis and two points symmetrically located on either side of the crack surfaces. The point A with coordinates (r_A, θ_A) is plotted on the upper surface and the point B $(r_B, -\theta_B)$ is plotted on the lower surface. The displacements u and v are defined as parallel to x and y, respectively. By considering the difference in displacements of the two faces of the crack, respectively v_A and v_B, the crack opening displacement (COD_{AB}) located at a distance $r_A = r_B$ of the crack tip can be calculated. For pure mode I crack, Eq. (4.2) can be rewritten in the following form:

$$CODy_{AB} = \frac{K_I}{G}\left[\sqrt{\frac{r_A}{2\pi}}\sin\frac{\theta_A}{2}\left[\frac{2}{1+\mu} - \cos^2\frac{\theta_A}{2}\right] - \sqrt{\frac{r_B}{2\pi}}\sin\frac{\theta_B}{2}\left[\frac{2}{1+\mu} - \cos^2\frac{\theta_B}{2}\right]\right] \tag{4.3}$$

and so the COD measured using DIC technique is substituted in Eq. (4.3) to obtain the mode I SIF value:

$$K_I = G \cdot CODy_{AB}\left[\sqrt{\frac{r_A}{2\pi}}\sin\frac{\theta_A}{2}\left[\frac{2}{1+\mu} - \cos^2\frac{\theta_A}{2}\right] - \sqrt{\frac{r_B}{2\pi}}\sin\frac{\theta_B}{2}\left[\frac{2}{1+\mu} - \cos^2\frac{\theta_B}{2}\right]\right]^{-1} \tag{4.4}$$

For points on the surface crack and sufficiently away from the crack tip (i.e. $r = r_A = r_B$), the coordinates of the points A and B can be considered as (r, π) and $(r, -\pi)$, respectively. In this way, Eq. (4.4) is simplified to:

$$K_I = \frac{E \cdot CODy_{AB}}{8}\sqrt{\frac{2\pi}{r}} \tag{4.5}$$

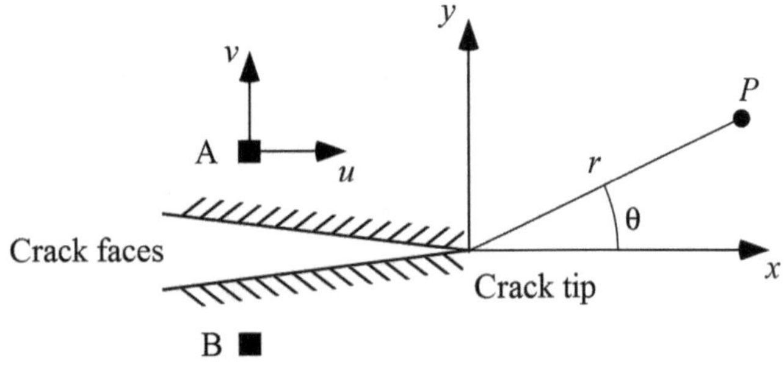

Fig. 4.1 Crack tip coordinate system

4.3 Calculating SIFs Using the Over-Deterministic Least-Squares Method

In the over-deterministic LSM approach, the Westergaard's stress function is used to determine the stress field ahead of the crack tip. This solution is written in the form of an infinite series expansion and it is not limited to the very near crack tip region. For the displacement components, the asymptotic equations are defined by [12]:

$$\begin{aligned} u = & \sum_{n=1}^{\infty} \frac{A_n}{2G} r^{n/2} \left\{ \kappa \cos\frac{n}{2}\theta - \frac{n}{2}\cos\left(\frac{n}{2}-2\right)\theta + \left\{\frac{n}{2} + (-1)^n\right\} \cos\frac{n}{2}\theta \right\} \\ & - \sum_{n=1}^{\infty} \frac{B_n}{2G} r^{n/2} \left\{ \kappa \sin\frac{n}{2}\theta - \frac{n}{2}\sin\left(\frac{n}{2}-2\right)\theta + \left\{\frac{n}{2} - (-1)^n\right\} \sin\frac{n}{2}\theta \right\} \end{aligned} \tag{4.6}$$

$$\begin{aligned} v = & \sum_{n=1}^{\infty} \frac{A_n}{2G} r^{n/2} \left\{ \kappa \sin\frac{n}{2}\theta + \frac{n}{2}\sin\left(\frac{n}{2}-2\right)\theta - \left\{\frac{n}{2} + (-1)^n\right\} \sin\frac{n}{2}\theta \right\} \\ & - \sum_{n=1}^{\infty} \frac{B_n}{2G} r^{n/2} \left\{ -\kappa \cos\frac{n}{2}\theta - \frac{n}{2}\cos\left(\frac{n}{2}-2\right)\theta + \left\{\frac{n}{2} - (-1)^n\right\} \cos\frac{n}{2}\theta \right\} \end{aligned} \tag{4.7}$$

Particularly, the first terms of the series A_I and B_I are related to the mode I and mode II as follows:

$$A_1 = \frac{K_I}{(2\pi)^{1/2}} \tag{4.8}$$

$$B_1 = \frac{K_{II}}{(2\pi)^{1/2}} \tag{4.9}$$

The displacements in Eqs. (4.6) and (4.7) are rewritten considering the possible rigid body motion (i.e. translation and rotation). When these terms are included, the crack tip displacement equations become:

$$u = \sum_{n=1}^{N} A_n f_{In}(r,\theta) - \sum_{n=1}^{N} B_n f_{IIn}(r,\theta) + T_x - Ry \tag{4.10}$$

$$v = \sum_{n=1}^{N} A_n g_{In}(r,\theta) - \sum_{n=1}^{N} B_n g_{IIn}(r,\theta) + T_y + Rx \tag{4.11}$$

where n is the number of terms of the series expansion of the displacement field, R is the rigid body rotation, and Tx and Ty are the components of the rigid body translation along x and y direction, respectively. The terms f_I, f_{II}, g_I and g_{II} are functions in terms of the polar coordinates r and θ.

Considering a set of m points collected from the region surrounding the crack tip, Eqs. (4.10) and (4.11) can be arrayed in the matrix form as:

$$[U] = [C] \cdot [\Delta] \tag{4.12}$$

where:

$$[U] = [u_1 \quad \cdots \quad u_m \quad v_1 \quad \cdots \quad v_m]^T \tag{4.13}$$

$$[\Delta] = [A_1 \quad \cdots \quad B_1 \quad \cdots \quad T_x \quad T_y \quad R]^T \tag{4.14}$$

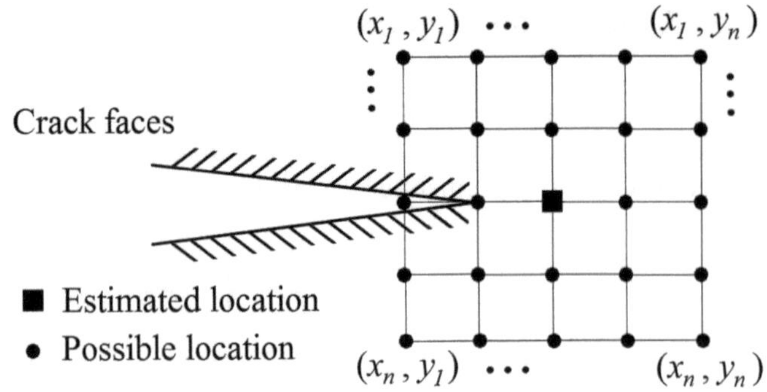

Fig. 4.2 Schematic representation of the grid used for crack tip location

$$[C] = \begin{bmatrix} f_{I1}(r_1,\theta_1) & \cdots & -f_{II1}(r_1,\theta_1) & \cdots & 1 & 0 & -y_1 \\ \vdots & \ddots & \vdots & \ddots & \vdots & \vdots & \vdots \\ f_{\mathrm{Im}}(r_M,\theta_M) & \cdots & -f_{II\mathrm{m}}(r_M,\theta_M) & \cdots & 1 & 0 & -y_m \\ g_{I1}(r_1,\theta_1) & \cdots & -g_{II1}(r_1,\theta_1) & \cdots & 0 & 1 & x_1 \\ \vdots & \ddots & \vdots & \ddots & \vdots & \vdots & \vdots \\ g_{\mathrm{Im}}(r_m,\theta_m) & \cdots & -g_{II\mathrm{m}}(r_m,\theta_m) & \cdots & 0 & 1 & x_m \end{bmatrix}^T \tag{4.15}$$

Therefore, by using the least squares method to solve Eq. (4.12), the unknown parameters are determined as:

$$[\Delta] = \left[[C]^T \cdot [C]\right]^{-1} \cdot [C]^T \cdot [U] \tag{4.16}$$

To extract the crack tip location, the following methodology is adopted. First, an estimate of the crack tip position is located using the DIC image at the maximum load applied in the loading cycle. Then, a square grid is placed centered at the crack tip estimated position, as shown in Fig. 4.2. The least square method is run repeatedly using all possible node-locations for the crack tip provided by the grid. The sum of squared errors (SSE) considering the experimental and modeling data at each location is calculated in order to find the best possible fit (i.e. the minimum error) and then the best estimation of the crack tip position. Once found the position on the grid that gives the minimum error, a new refined grid can be generated on this position to obtain a more accurate location for the crack tip.

After determining the crack-tip location, the data points located in the region immediately ahead of the crack tip are eliminated to avoid the intensively nonlinear zone around the crack tip and also to avoid three dimensional effects. The size of the region used for analysis and the desired accuracy determines the total number of terms adopted in the series expansion to adequately describe the stress state. According to [5, 10], the SIF value obtained from the over-deterministic least-squares method converge to a constant value as the number of terms increases.

4.4 Calculating SIFs Using the J-Integral Formulation

The J-integral is defined as a path-independent line integral that quantifies the strength of the singular stresses and strains on a contour surrounding a crack tip. Consider an arbitrary path (Γ) around the crack tip (see Fig. 4.3), Rice [13] defined the J-integral as:

$$J = \int_\Gamma W dy - \vec{T} \cdot \frac{\delta \vec{u}}{\delta x} ds \tag{4.17}$$

where Γ is the integration path around the crack tip which is evaluated in a counterclockwise direction beginning on the lower crack surface and ending on any point on the upper crack surface, ds is the length increment along the contour Γ, T is the traction vector defined by the normal to Γ in the outside direction, and W is the total strain energy density:

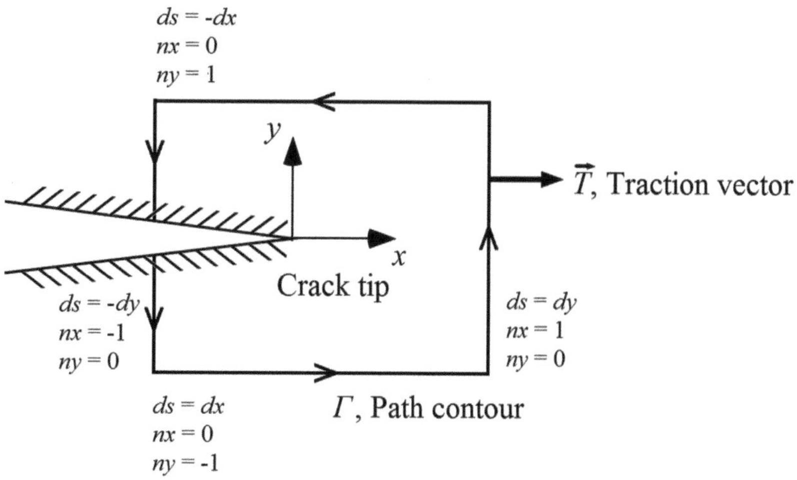

Fig. 4.3 Rectangular path for J calculation

$$W = \int_0^{\varepsilon_{ij}} \sigma_{ij} d\varepsilon_{ij} \tag{4.18}$$

The numerical value of the J-integral can be related to the SIF or to the pseudo-SIF. For plane stress, under Mode I loading conditions, this relation is

$$J = K^2/E \tag{4.19}$$

The experimental procedure to calculate the J-integral is based on the direct evaluation of the line integral defined in Eq. (4.17). In this way, the J-formulation can be expressed explicitly as:

$$J = \int_\Gamma \left(W - \sigma_{xx}\frac{\delta u}{\delta x} - \sigma_{xy}\frac{\delta v}{\delta x} \right) dy + \int_\Gamma \left(\sigma_{xy}\frac{\delta u}{\delta x} + \sigma_{yy}\frac{\delta v}{\delta x} \right) dx \tag{4.20}$$

with

$$W = [(\sigma_1 \cdot \varepsilon_1) + (\sigma_2 \cdot \varepsilon_2)]/2 \tag{4.21}$$

Using strain-displacement equations and generalized Hooke's law (considering linear elastic conditions prevailing along the Γ integration path), the stresses and strains of the Eqs. (4.20) and (4.21) can be computed using the displacement gradients obtained from the raw DIC data.

The procedure to calculate the four displacement gradients is shown in Fig. 4.4. In this example, a circular path was used. Once a distance Δx is defined, the path is shifted by a distance of $\Delta x/2$ in the positive and negative x-directions to calculate the partial derivatives of the displacement vector $\delta u/\delta x$ and $\delta v/\delta x$ using

$$\frac{\delta u_i}{\delta x} = \frac{u_i'' - u_i'}{\Delta x} \tag{4.22}$$

$$\frac{\delta v_i}{\delta x} = \frac{v_i'' - v_i'}{\Delta x} \tag{4.23}$$

Similarly, the procedure is applied to calculate the corresponding derivatives $\delta u/\delta y$ and $\delta v/\delta y$ using a shifted distance Δy. It is recommended that the distances Δx and Δy to be approximately 1 % of the total length of the path [14].

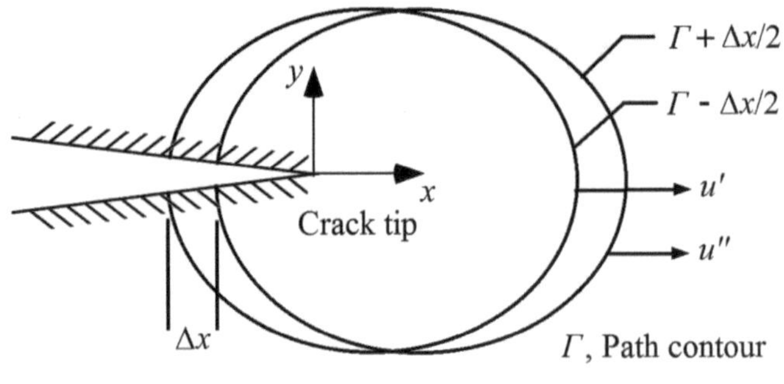

Fig. 4.4 Procedure to calculate the derivatives of the displacement vector [14]

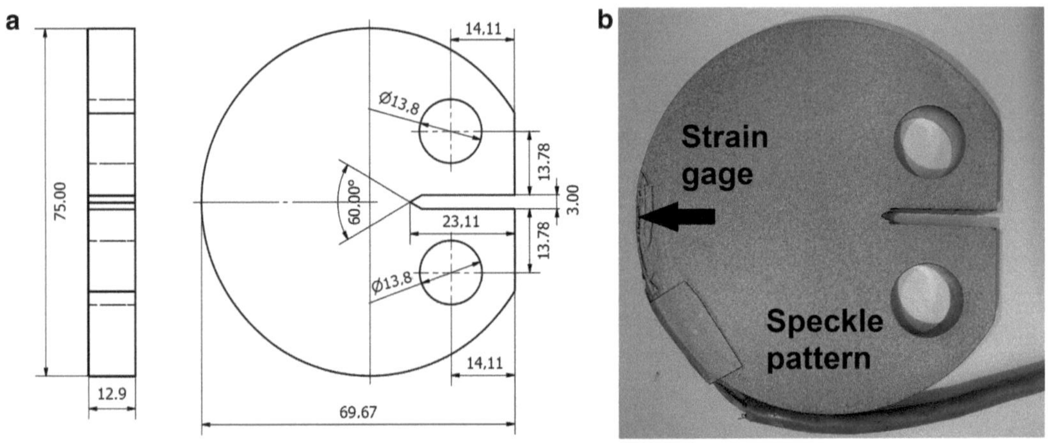

Fig. 4.5 (**a**) Geometry and dimensions of the specimen, (**b**) speckle pattern applied to the sample and location of the strain gage used in the test

4.5 Experimental Procedure

The three methodologies for determining Mode I SIF were applied to data obtained from fatigue crack growth tests on disk-shaped compact-tension specimens (according with ASTM E399) machined from an AISI 4340 steel 3″ diameter bar, heat treated to tempered martensite with 200HV-10. The geometry and dimensions of the specimen are depicted in Fig. 4.5a. For DIC measurements, the specimen was first coated with a thin layer of a white paint on its surface, and then black dots were spray-painted on the white background (see Fig. 4.5b). In addition, a uniaxial strain gage (Kyowa Electronic Instruments, KFG-5-120C1-11) was bonded on the back face of the specimen to collect strain information during testing (see Fig. 4.5b).

The fatigue test was performed in a 100 kN servo-hydraulic machine Instron model 8501. The images were recorded by a stereoscopic system composed by two 5-MP CCD cameras (Point Grey GRAS-50S5M) equipped with 35-mm lenses. A traveling microscope was placed in front of the unpainted surface of the specimen to obtain crack length measurements, as shown in Fig. 4.6.

The specimen was pre-cracked in accordance with the standard ASTM E647 using a stress ratio of $R = 0.1$ at 10 Hz until a crack with length of approximately 0.1 mm initiated. Then, the specimen was loaded at 20 Hz under a quasi-constant SIF range $\Delta K = 31$ MPa$\sqrt{\text{m}}$ and $R = 0.1$. No overloads or similar events were applied on the tested specimen. The nominal SIF values were calculated using the standard ASTM E399 equation:

$$K_I = \frac{P}{t\sqrt{w}} \cdot \frac{(2 + a/w)}{(1 - a/w)^{1.5}} \cdot \left[0.76 + 4.64 \cdot a/w - 11.58 \cdot (a/w)^2 + 11.43 \cdot (a/w)^3 - 4.08 \cdot (a/w)^4\right] \quad (4.24)$$

where w and t are the specimen width and thickness, a is the crack length (measured from the load line), and P is the applied load.

DIC data were gathered when the fatigue crack reached a total length of 6 mm, using the loading history shown in Fig. 4.7. The loading velocity was reduced to 0.05 kN/s so that the DIC images could be taken. The recorded loading history

Fig. 4.6 Experimental set up used for validation

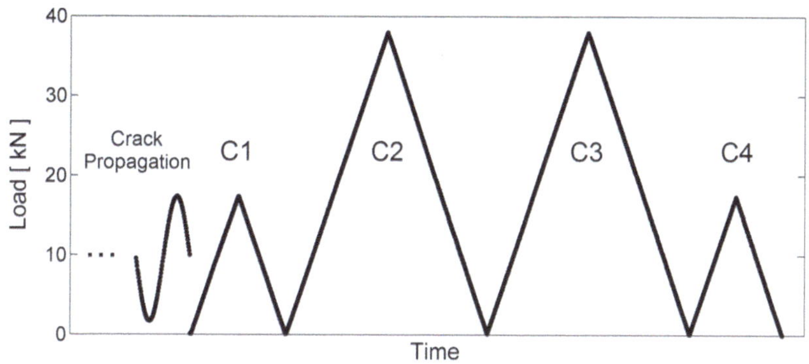

Fig. 4.7 Load history applied to specimen

consisted of one cycle, C1, with a peak load value equal to 17.4 kN similar to the previous loading; two cycles, C2 and C3, with peak loads of 37.8 kN, and finally one cycle, C4, similar to C1. The load cell signal was synchronized with the cameras, with an image captured at every 0.2 kN load increment. The DIC analysis was performed using the commercial software VIC-Snap and VIC-3D [15] with a subset size of 31 pixels, step size of 7 pixels and strain windows of 15.

4.6 Results

The first step in the DIC analysis is to find the crack tip location using the methodology described in Sect. 4.3 based on the LSM approach, using the time-step images related to the maximum load applied in cycle C_2. The search algorithm used a square grid with size 3 $\times$ 3 mm and step space of 0.05 mm. The location obtained was confirmed after removing the paint used for the DIC.

Figure 4.8a shows the dimensions of the field of view used in the optical setup. Figure 4.8b shows two pair of symmetrical points located far away from the crack tip, selected to obtain the SIF values based on the COD measurements. The two pair of points (A_1, B_1) and (A_2, B_2) were located at 3.3 and 4.0 mm from the tip, respectively. Figure 4.8c shows the data points used on the LSM approach without points located near the crack tip and crack surfaces. The plastic zone size around the crack tip was estimated to be approximately 1 mm using Irwin's plane stress estimative based on the peak C1 load. Data points inside this area, ahead of the crack tip, were excluded from the DIC analyses. The SIF values for this approach were obtained using ten parameters in the series solution for the displacement fields (Eqs. (4.6) and (4.7)). Figure 4.8d shows the contour path around the crack tip used in the J-integral. The total path length was 16 mm and the line path was approximately 3.5 mm far ahead of the crack tip.

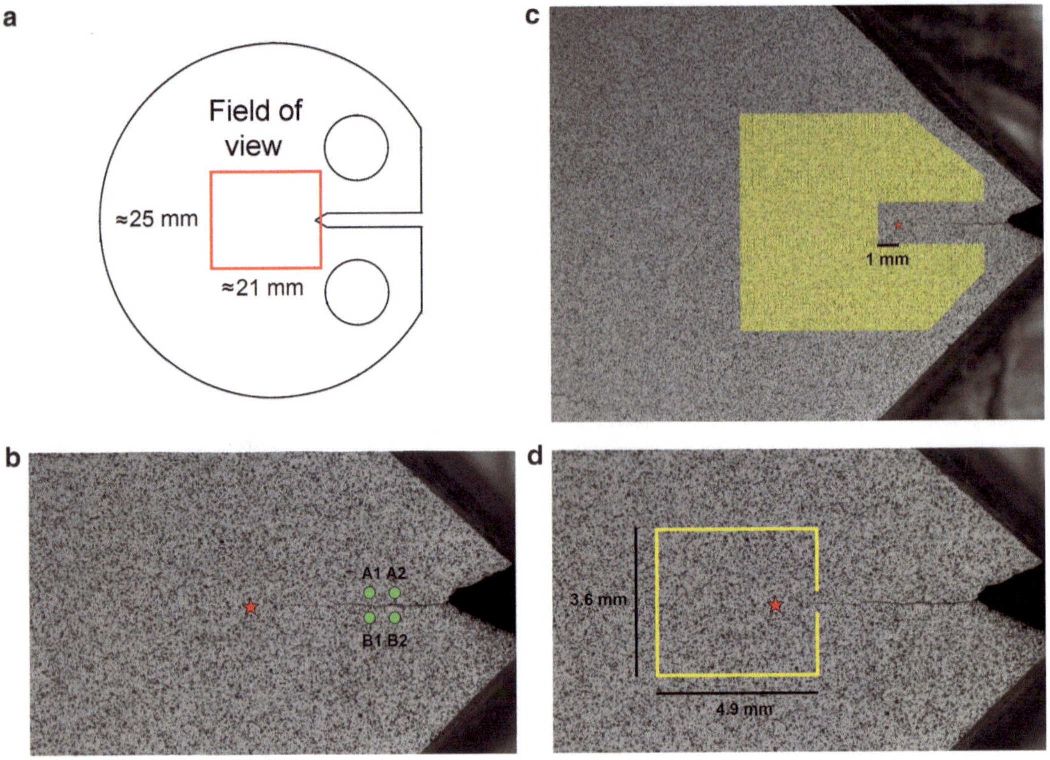

Fig. 4.8 (**a**) Field of view used for DIC analysis. (**b**) Symmetrical points used in COD approach. (**c**) Data points used in over-determinist LSM approach with an exclusion area of 1 mm. (**d**) Rectangular path used in J-integral approach

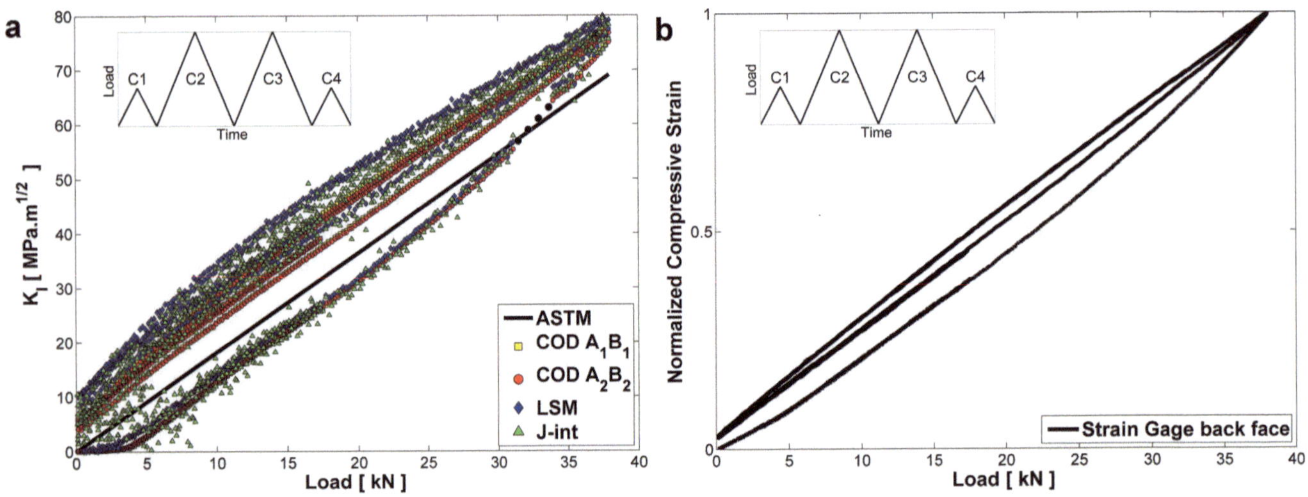

Fig. 4.9 (**a**) SIF measurements from the three experimental approaches (more enhanced and clearer plots for the individual cycles C1–C4 are presented in Figs. 4.10, 4.11, 4.12, 4.13 and 4.14), (**b**) strain distribution from strain gage placed on the back face of the DC(T) specimen

Results of the analysis performed for the complete load history (cycles C1 to C4) are shown in Fig. 4.9. Figure 4.9a shows the SIF evaluation using the three LEFM approaches described previously, neglecting the non-linear effects induced by crack closure and by additional plastification around the crack tip due to the peak load increase in cycles C2 and C3. The DIC analysis first considered the image captured at beginning of cycle C1 (unloaded stage) as the reference image. From the curves showed in Fig. 4.9a, it is clear that the K_I results obtained from the three methodologies have the same trend, a fact confirmed by the strain gage positioned in the back face of the specimen, see Fig. 4.9b. It can be also seen that the linear part of experimentally determined K_I results follow the same trend showed by the ASTM Eq. (4.24).

Figure 4.9 clearly shows that there are some non-linear phenomena that affect the measured SIF, namely crack closure at low loads as well as plasticity ahead of the crack tip and crack tip blunting at higher loads in cycle C2, which are not taken into account in Eq. (4.24) and in the equations used to calculate K_I from the DIC measurements, all of which assume a linear elastic (LE) behavior. Moreover, the coordinates of the crack tip used in Eqs. (4.4), (4.5), (4.16), and (4.24) assume that the crack is fully opened and does not move during the load cycle. However, Figs. 4.9a and 4.10 show that the crack remains partially closed during cycle C1 until it opens at about 4.5 kN. A detailed analysis to determine the peeling off behavior observed during the partial opening of the crack by finding the "operational crack tip location" using the procedure described in Sect. 4.3 and Fig. 4.2, when the load value varies from its minimum value up to the opening load at 4.5 kN, is out of the scope of the present paper and will be presented elsewhere. Anyway, as depicted in Figs. 4.10, 4.11, 4.12, 4.13 and 4.14, the loading and unloading stages of the K_I curves follow parallel to the ASTM Eq. (4.24) curve when the crack is fully opened and is not suffering additional plastification during cycle C2. Therefore, it is recommended that the K_I results presented in Figs. 4.9, 4.10, 4.11, 4.12, 4.13 and 4.14 be interpreted as pseudo-K_I or pseudo-SIF values calculated assuming an LE behavior, which generate meaningful K_I values only from the slopes of the linear parts of the presented curves.

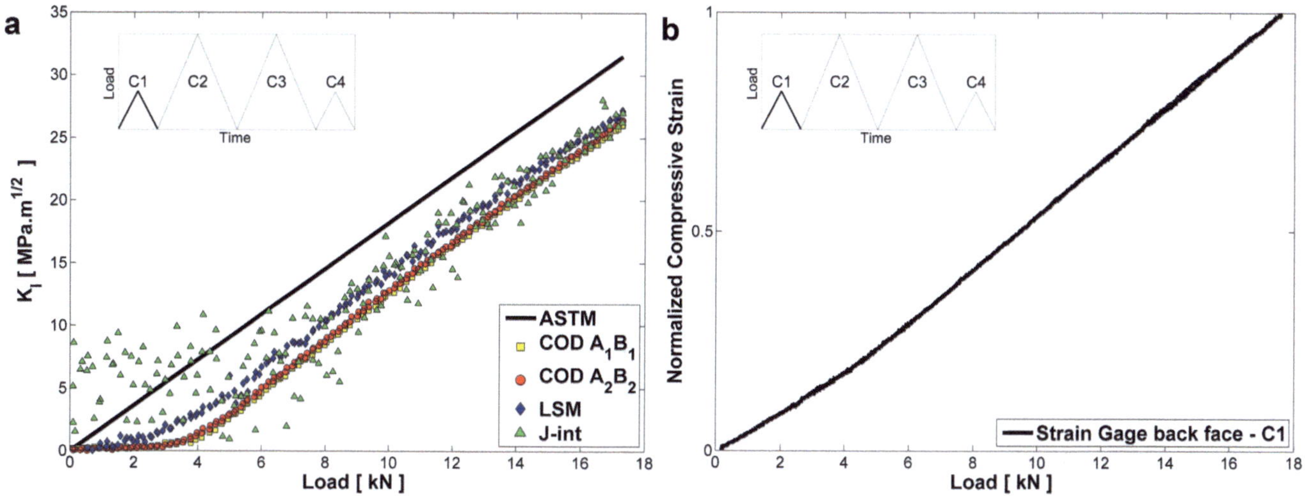

Fig. 4.10 SIF evaluation for cycle C1 (**a**) measurements from the three experimental approaches, (**b**) strain distribution from strain gage placed on the back face of the DC(T) specimen

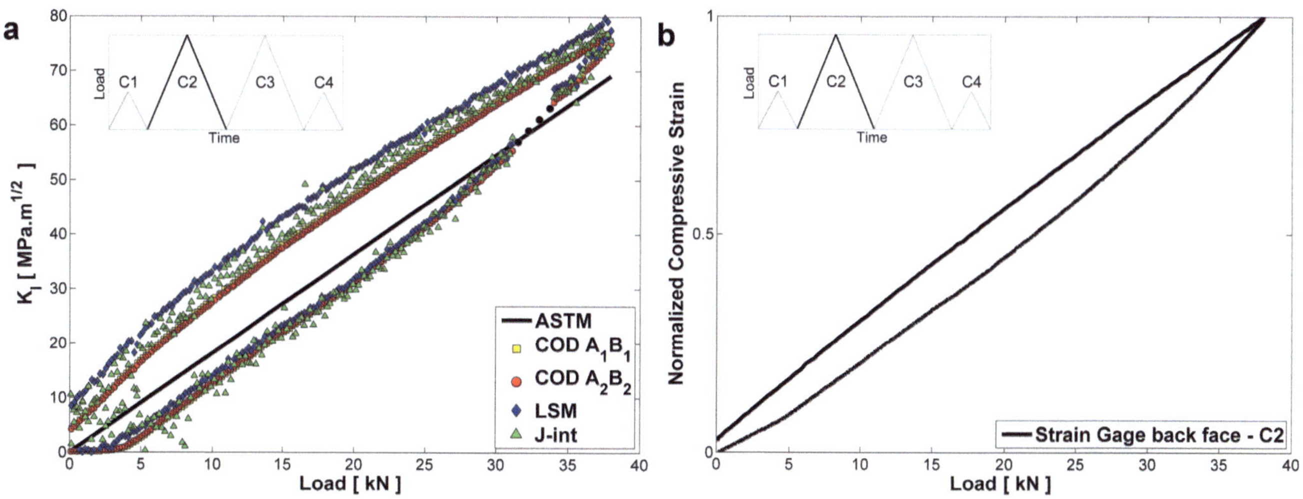

Fig. 4.11 SIF evaluation for cycle C2 (**a**) measurements from the three experimental approaches, (**b**) strain distribution from strain gage placed on the back face of the DC(T) specimen

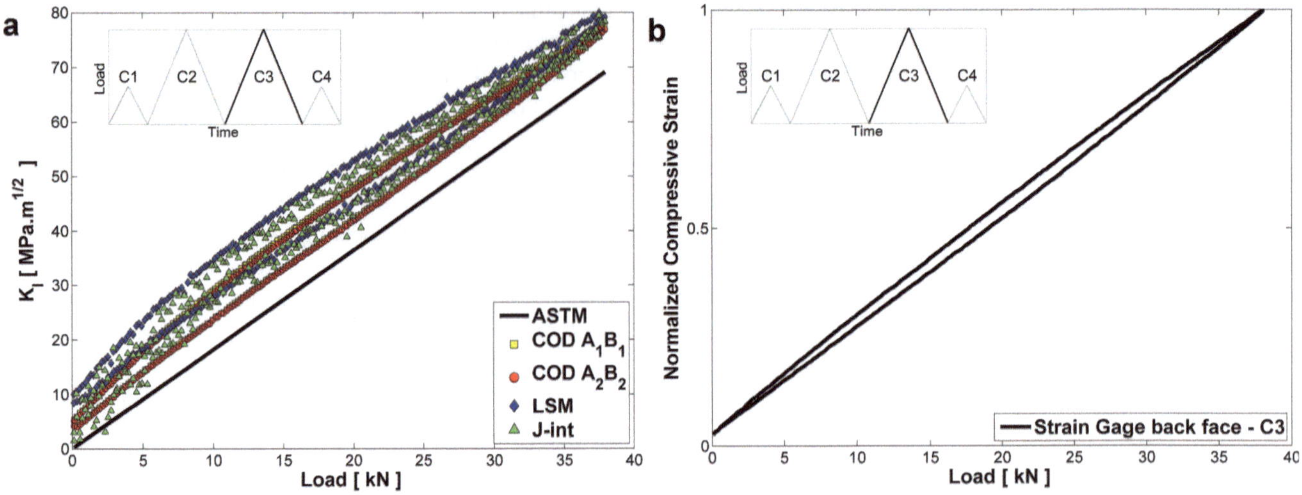

Fig. 4.12 SIF evaluation for cycle C3 (a) measurements from the three experimental approaches, (b) strain distribution from strain gage placed on the back face of the DC(T) specimen

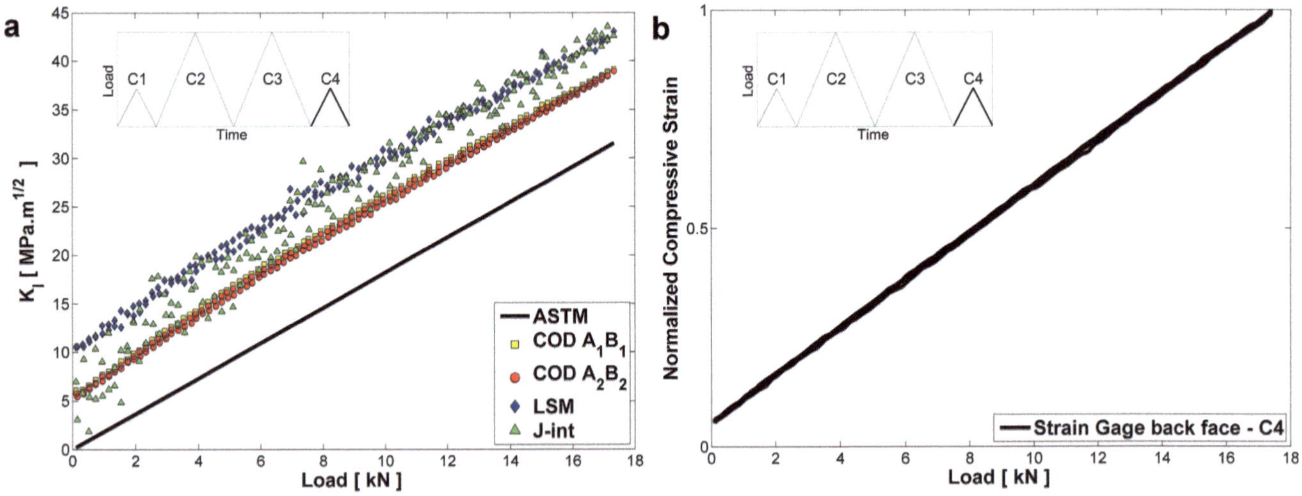

Fig. 4.13 SIF evaluation for cycle C4 (a) measurements from the three experimental approaches, (b) strain distribution from strain gage placed on the back face of the DC(T) specimen

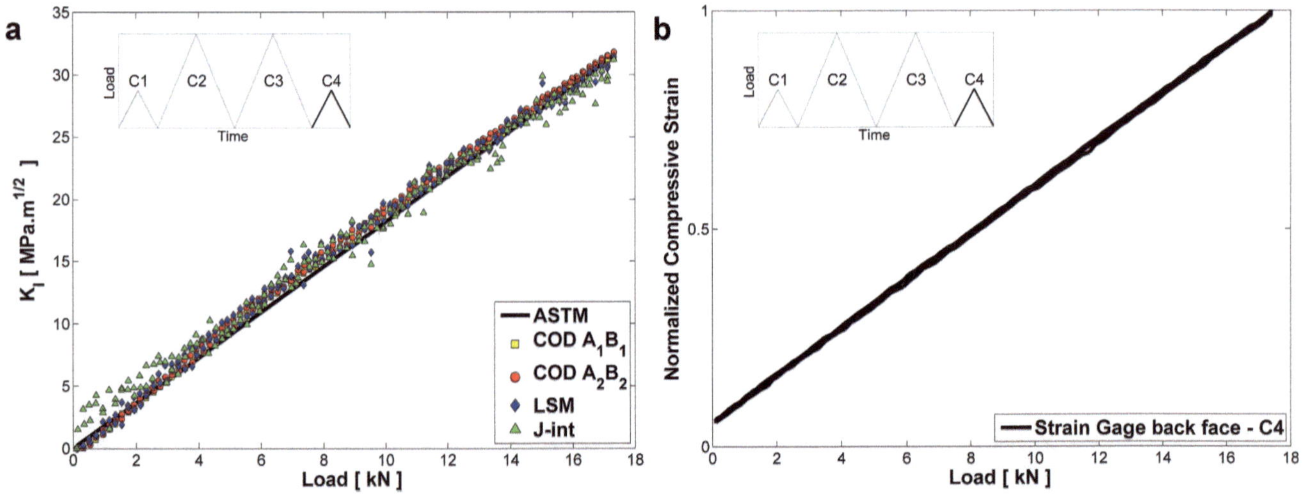

Fig. 4.14 Measurements from the three experimental approaches analyzed with the reference image taken at the beginning of the cycle C4

Figure 4.10 shows in details the pseudo-K_I values for loading cycle C1, before the C2 overload. It can be seen that the SIFs LE-calculated in Fig. 4.10a and the strains measured in Fig. 4.10b are clearly affected by crack closure, which causes an offset between the theoretical and the "experimental" SIF values. Only after the crack is fully opened the pseudo-K_I curve remains parallel to the ASTM Eq. (4.24) curve up to the maximum load value reached in the previous cycles.

Figures 4.11 and 4.12 show results for C2 and C3 load cycles. Note that once again both the DIC analyses and the strain gage measurements have the same behavior, a clear indication of their compatibility. Note as well that the loading part of the C2 curves follows the same path of the C1 cycle until the C1 peak load is reached, including the initial closure behavior. However, from this point and on, both the pseudo-K_I and the strain gage curves show another non-linear behavior. Moreover, all unloading C2 curves present a highly hysteretic behavior and a significant zero-shift, detected by the three DIC SIF calculation approaches and by the strain gage, which induces a permanent pseudo-K_I.

It is important to emphasize that the zero-shift induced by the C2 overload is not observed again in cycle C3, which is identical to C2 and in particular reaches the same load peak as depicted in Fig. 4.12. Note as well that the C3 unloading curves remain almost parallel to the Eq. (4.24) curve, a clear indication of the stable behavior expected after the material ahead of the crack tip has been further plastified by C2.

The results of the SIF evaluation for the cycle C4 after the overload are shown in Fig. 4.13. This experiment indicates that after the overload, the crack tip is being held open even at zero loading. In this way, the effects of the crack closure were eliminated, as shown in Fig. 4.13a. The same behavior is exhibited by the strain gage readings shown in Fig. 4.13b. This behavior is compatible with crack tip blunting induced by the higher C2 and C3 cycles. In addition, a new DIC analysis was performed by selecting the image at the beginning of the cycle C4, with the specimen in the unloaded stage, as the reference image. Therefore, the residual strains and stresses caused by the overload are not taken into account in the new analysis. As it is shown in Fig. 4.14, the SIF measurements obtained after the overload agreed very well with the analytical LE solution, which does not consider the nonlinearities existing in the crack tip region.

4.7 Conclusions

This work addresses the issues involved in LE SIF measurements in actual fatigue cracks when they present, as usual, non-linear behaviors such as crack closure at low load levels and crack tip plasticity and blunting after overloads. Albeit the LE analyses clearly reproduce the expected behavior when the crack is fully opened and it is not presenting significant further plastification, it is clear that the blind use of LE equations to calculate the SIF values from the DIC measurements is not an acceptable or even reasonable procedure, since it can lead to quite improper conclusions while these non-linear effects are affecting the crack behavior. This point is not new, and Elber himself used it to identify crack closure in his pioneer measurements [16], but it has been sometimes neglected in contemporary literature. These non-linear effects cannot of course be analyzed by LE equations, and must be considered by appropriate procedures.

References

1. Sutton, M.A., McNeill, S.R., Helm, J.D., Chao, Y.J.: Advances in Two-Dimensional and Three-Dimensional Computer Vision, Photomechanics, pp. 323–372. Springer, Berlin (2000)
2. Sutton, M.A., Orteu, J.J., Schreier, H.: Image Correlation for Shape, Motion and Deformation Measurements: Basic Concepts, Theory and Applications. Springer, New York (2009)
3. McNeill, S., Peters, W., Sutton, M.: Estimation of stress intensity factor by digital image correlation. Eng. Fract. Mech. **28**(1), 101–112 (1987)
4. Rethore, J., Gravouil, A., Morestin, F., Combescure, A.: Estimation of mixed-mode stress intensity factors using digital image correlation and an interaction integral. Int. J. Fract. **132**(1), 65–79 (2005)
5. Yoneyama, S., Morimoto, Y., Takashi, M.: Automatic evaluation of mixed-mode stress intensity factors utilizing digital image correlation. Strain **42**(1), 21–29 (2006)
6. Ju, S.-H., Liu, S., Liu, K.: Measurement of stress intensity factors by digital camera. Int. J. Solids Struct. **43**(5), 1009–1022 (2006)
7. Lopez-Crespo, P., Shterenlikht, A., Patterson, E., Yates, J., Withers, P.: The stress intensity of mixed mode cracks determined by digital image correlation. J. Strain Anal. Eng. Des. **43**(8), 769–780 (2008)
8. Zhang, R., He, L.: Measurement of mixed-mode stress intensity factors using digital image correlation method. Opt. Lasers Eng. **50**(7), 1001–1007 (2012)
9. Becker, T., Mostafavi, M., Tait, R., Marrow, T.: An approach to calculate the J-integral by digital image correlation displacement field measurement. Fatigue Fract. Eng. Mater. Struct. **35**(10), 971–984 (2012)
10. Harilal, R., Vyasarayani, C., Ramji, M.: A linear least squares approach for evaluation of crack tip stress field parameters using DIC. Opt. Lasers Eng. **75**, 95–102 (2015)

11. Sanford, R.J.: Principles of Fracture Mechanics. Pearson College Division (2003)
12. Atluri, S.N., Kobayashi, A.: Chapter 1, Mechanical responses of material, Handbook on experimental mechanics, pp. 1–40 (1987)
13. Rice, J.R.: A path independent integral and the approximate analysis of strain concentration by notches and cracks. J. Appl. Mech. **35**(2), 379–386 (1968)
14. ANSYS online manual, Structural Analysis Guide. http://www.ansys.stuba.sk/html/guide_55/g-str/GSTR10.htm
15. Correlated Solutions Inc., U. S. A. http://www.correlatedsolutions.com/
16. Elber, W.: The Significance of Fatigue Crack Closure, Damage Tolerance in Aircraft Structures. ASTM International, West Conshohocken (1971)

Chapter 5
Characterization of Sub-surface Damage During the Early Stage of Stress Corrosion Cracking by Nano Indentation

Denizhan Yavas, Pratyush Mishra, Ashraf F. Bastawros, Kurt R. Hebert, and Pranav Shrotriya

Abstract Stress Corrosion Cracking (SCC) is a cracking process observed in metals under tensile stress and corrosive environment. It has been reported as one of the major failure modes in high strength steel pipelines over few decades. This study aims to characterize sub-surface damage, which leads to near surface mechanical property changes during the initiation stage of SCC. The quasi-static and dynamic nano indentation technique, were utilized to resolve the mechanical property variation in the near-surface region. The measurements indicated significant softening and modulus reduction near the grain boundaries. Such findings could further elucidate the underlying deformation mechanisms during the early stage of SCC as a function of degradation level.

Keywords Stress corrosion cracking • Pipeline steels • Sub-surface damage • Non-destructive evaluation • Nano indentation

5.1 Introduction

Stress corrosion cracking (SCC) refers to the failure by cracking of normally ductile metals in a corrosive environment, when they are subjected to tensile stress. SCC affects a wide range of high strength alloys. For example 21 % of steel pipeline failures are attributable to SCC [1]. Typically, SCC initiates at localized corrosion sites such as pits and grain boundaries and is greatly affected by the environment. The critical aspect of SCC is the brittle failure of the ductile high-strength alloys, and the associated reduction of the fracture toughness. There is widespread debate about the mechanism of SCC, including the contributions of mechanical vs. chemical processes in the initial damage percolation, leading to a macroscopic crack growth and the associated role of hydrogen embrittlement. Several mechanisms suggest that brittle cracking is produced by localized plastic deformation activities caused by dislocation emission and motion accompanying localized corrosion at the crack tip [2–4]. The pH plays a significant role in defining the operative chemo-mechanical mechanisms that leads to SCC in steels. High pH SCC is attributed to anodic dissolution at the grain boundaries and repeated rupture of passive films that form over the crack tip, and thereby leading to intergranular failure. The near-neutral pH stress corrosion cracks is due to the synergistic effects of stress and hydrogen on anodic dissolution at crack-tip, and thereby leading to transgranular failure [5, 6].

Pre-emptive detection and control of SCC is difficult owing to its prolonged incubation stage [7], in which the corrosion process is first established locally and then cracks propagate at a very low rate, for the evolution of damage depth. This incubation period may be followed by catastrophic failure due to unstable crack growth. Current NDE methods in the laboratory are mastering the detection of defects at scales of tens to hundreds of microns and embrace the "find and fix" philosophy. There are critical needs to understand these degradation mechanisms at the percolation stage at the nano- to micron-length scales, and to identify much earlier the precursor signals associated with the initiation stage of cracking.

D. Yavas
Department of Aerospace Engineering, Iowa State University, Ames, IA 50011-2271, USA

P. Mishra • K.R. Hebert
Department of Chemical and Biological Engineering, Iowa State University, Ames, IA 50011-2271, USA

A.F. Bastawros (✉)
Department of Aerospace Engineering, Iowa State University, Ames, IA 50011-2271, USA

Department of Mechanical Engineering, Iowa State University, Ames, IA 50011-2271, USA
e-mail: bastaw@iastate.edu

P. Shrotriya
Department of Mechanical Engineering, Iowa State University, Ames, IA 50011-2271, USA

Y. Zhu, A.T. Zehnder (eds.), *Experimental and Applied Mechanics, Volume 4*,
Conference Proceedings of the Society for Experimental Mechanics Series, DOI 10.1007/978-3-319-42028-8_5

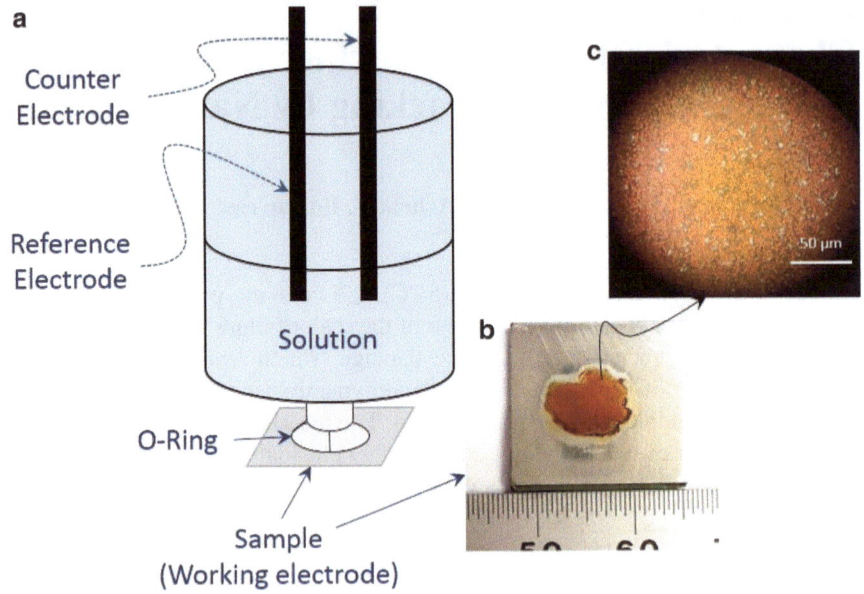

Fig. 5.1 (**a**) Schematic drawing of the electrochemical cell used in the corrosion experiments, (**b**) Optical image of the corroded sample, (**c**) Optical micrograph of the corroded surface

Accordingly, the scope of this paper is to identify the microstructural and material property changes accompanying the initiation stage of SCC. All measurements will be carried out by quasi-static and dynamic nanoindentation on an accelerated corrosion experiment of a typical pipeline steel (X70) exposed to an accelerated corrosion process. Measurement will be carried out on the corroded surface with its chemical product intact, as well as a light cleaning to assess the subsurface state and mechanical degradation.

5.2 Electrochemical Experiment and Sample Preparation

Electrochemical experiments were performed to mimic and accelerate the environmentally occurring corrosion processes in X70 pipeline steel. A conventional three-electrode cell was used to perform the corrosion experiments, with the samples as the working electrode, platinum wire as the counter electrode and Ag/AgCl as the reference electrode (Fig. 5.1a). The samples were cut from the walls of a steel pipe into sample dimensions of 18 mm × 18 mm × 2 mm as shown in Fig. 5.1b. The working surface of the samples was polished on a 400, 600, and 1000-grit grinding papers, respectively, then cleaned by deionized water and ethanol prior to the electrochemical experiments. Bicarbonate solution (1 M $NaHCO_3$) was selected as the electrolyte, which was prepared from analytical grade chemicals and ultra-pure water (18.2 MΩ cm). The average pH range was 8.4–8.8. The first step of the electrochemical experiments was to obtain the polarization (potential-current) curves to determine the susceptible range of voltage to attain a peak current, commensurate of maximum level of SCC in X-70 pipeline steel. The polarization experiment consists of two steps: (i) cathodic polarization at −1 V/SCE for 5 min to remove any native oxide layer formed on the surface of the samples and, (ii) anodic polarization at 0.1 mV/SCE. For the examined X70 steel sample and corrosive solution, the maximum current density was observed at the applied voltage range between −0.6 and 0.5 V/SCE, which is in good agreement with reported range in the literature [8]. The second set of experiments at a constant applied voltage of −0.5 V/SCE were conducted with variable exposure time, ranging between 10 and 120 min to examine the role of the corrosion time on the surface evolution and the subsurface damage.

5.3 Surface Morphology Characterization

The corrosion product layer deposited on the surface of the sample gave a pale green color. The color eventually turned to light brownish color after the removal of the samples from the solution as shown in Fig. 5.1b, c. It was previously proposed that the corrosion products consist of hydroxide and carbonates [9]. Figure 5.2 shows the corrosion product formation for 30 min and 120 min corroded samples, respectively. The corrosion products form semi-spherical-like features on the surface.

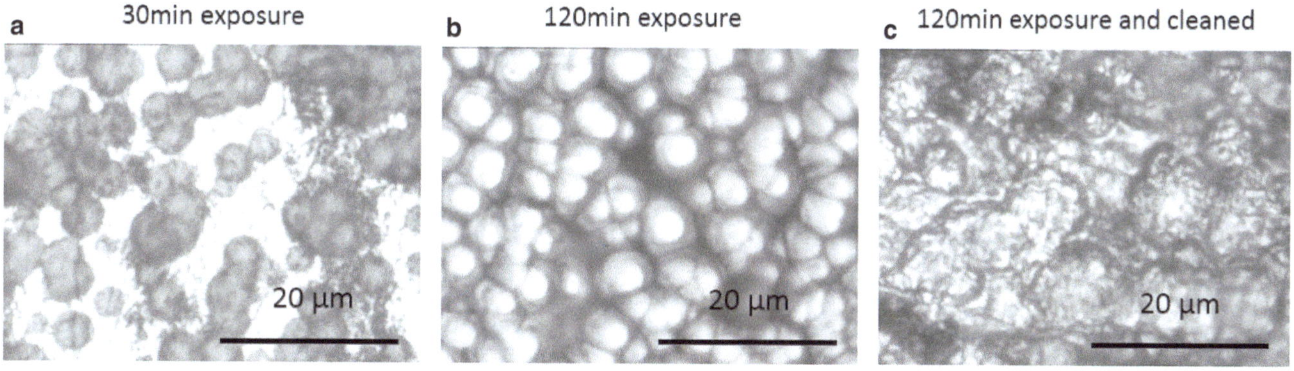

Fig. 5.2 Optical micrographs of the corroded surfaces at different exposure time, under constant peak potential of −0.5 V/SCE in bicarbonate solution (1 M $NaHCO_3$) (**a**) 30 min, and (**b**) 120 min, and (**c**) 120 min corroded surface after removal of corrosion products in acetic acid for 8 min

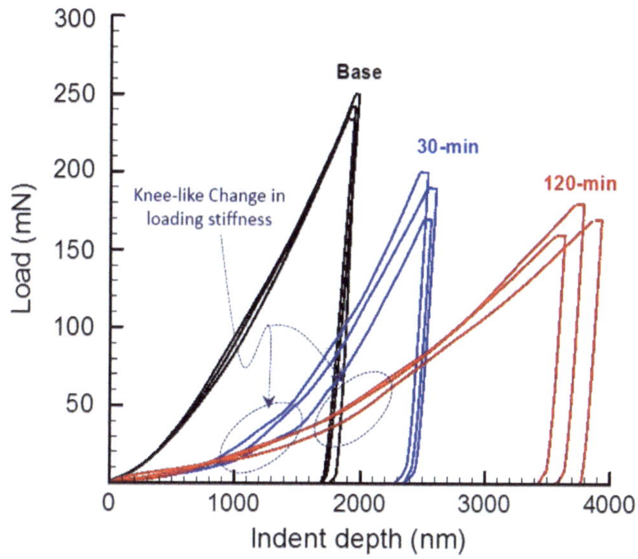

Fig. 5.3 A representative set of force-indentation depth curves for the base (no corrosion), 30 min and 120 min corroded samples

It should be also noted that the corrosion product on the surface become much denser and thicker, as the exposure time increases. The thickness of the corrosion product was measured by a Zygo 6200 NewView optical interferomter to be around 2 µm for 30 min exposure, as it was not fully covered by the corrosion product. Figure 5.2c is an optical image of the surface after cleaning by a diluted (0.1 M) acetic acid solution to remove the corrosion products. Grain boundaries (GB) became well defined with grooving in the range of 0.1–0.3 µm (darker regions). The average grain size is in the range of 5–10 µm (lighter regions).

5.4 Characterization of Corrosion Product Layer

The mechanical properties of the corrosion product deposited on the sample surface were measured by high load quasi-static indentations. Hysitron TI 950 TriboIndenter with the high load transducer (load range 0–500 mN) was utilized under force control mode. A trapezoidal loading profile, which consists of a 5 s linear loading, 2 s hold at the peak load, and 5 s linear unload, was applied to impose the regular patterns of indents using a Berkovich indenter tip. A sequence of indentations were performed on each sample with progressive increase of the peak load between 1 and 250 mN. Such loading range resulted in a maximum indentation depth of 4 µm. Figure 5.3 depicts a representative set of force- indentation depth curves for the base (no corrosion), 30 and 120 min corroded samples. One can clearly note a transition from softer to a higher stiffness response or "a knee-like" change at a characteristic indentation depth of about 1 µm for 30 min exposure time, and

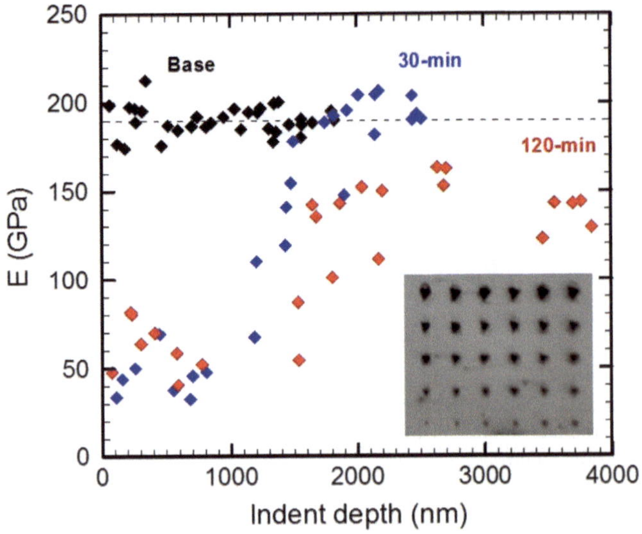

Fig. 5.4 Indentation modulus as a function of indent depth for the product layer, showing the extent of the corrosion products. The product layer has a modulus of about 30 % the base steel. The insert is an optical image of the progressive indent imprints on the base material

2 μm for 120 min exposure time. Such knee-like behavior or change of indentation stiffness can be attributed to a transition from the soft corrosion product layer to bulk base material. Figure 5.4 shows the variation of the indentation modulus E (derived from the contact analysis by Oliver and Pharr [10]) as a function of the indent depth for the examined three cases. For the base sample, E is around 190 GPa, and is independent of the indentation depth. Though, a noticeable drop on E can be seen for the corroded samples with chemical products. The corrosion product layer has a modulus of about 25 % that of the base material (E ~ 60 GPa) with a depth of ~1200 nm for 30 min exposure and ~2000 nm or more for the 120 min exposure. The measured E of the corrosion product is about the 30 % that of the bulk. The measurements were carried out in the dry state. It remains to be seen if the hydroxylated state might exhibit the same stiffness, as such layer may contribute to the balance of the residual stresses developed during the corrosion process. An in situ measurements are being carried out to elucidate such possibility.

5.5 Near-Surface Damage Evolution

Characterization of the near surface damage was conducted by high resolution dynamic nanoindentation technique employing Hysitron nanoDMA® III transducer with 90° cube corner indenter tip. The load profile is similar to the one used in quasi-static indentation test, though a small oscillatory force was simultaneously superimposed during the loading step, which facilitates continuous measurement of the material response. The first set of experiment were performed on the 30 min exposure time. Figure 5.5 shows the force-indentation depth curves for the base material and 30 min exposure time. At an early stage of the deformation history, displacement bursts were typically observed, which corresponds to individual slip activities and dislocation bursts. At the same displacement level, the observed nucleation load for such displacement burst is relatively lower by ~45 % for the 30 min exposure, compared to the reference (no corrosion) surface. The critical load for the displacement burst can be associated with the Peierls-Nabarro stress through the following expression [11],

$$\tau_{PN} \approx \tau_{\max} = 0.31\left(\frac{6E^2}{\pi^3 R^2}P_{cr}\right)^{1/3}, \tag{5.1}$$

where τ_{PN} is the Peierls-Nabarro stress, which is the critical stress required for dislocation activity, $\tau_{\max}$ is the maximum shear stress underneath the spherical indenter tip, E is the indentation modulus of the material, R is the radius of the indenter tip, and P_{cr} is the critical force required for dislocation activity.

The corrosion process has resulted in a reduction of the lattice resistance to dislocation mobility by about 25 %. It could be argued that such reduction is an outcome of the accumulation of chemically enhanced nucleation and diffusion of point defects (e.g. hydrogen entrapment and lattice level vacancies).

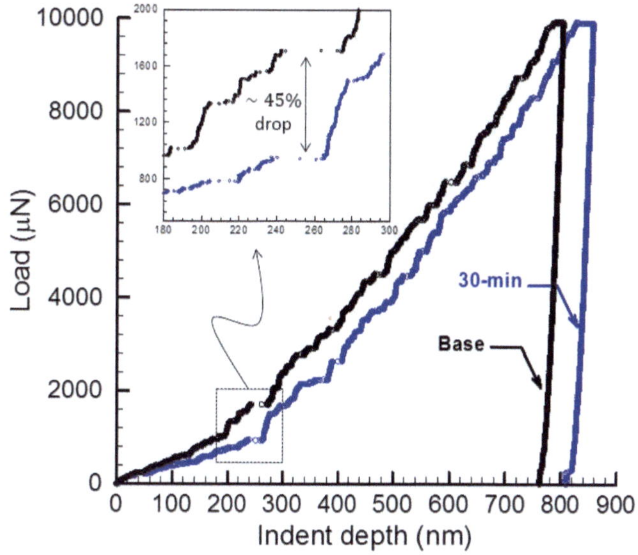

Fig. 5.5 Force-indentation depth curves for the base (no corrosion) material surface, and 30 min corroded sample surface, showing changes of the nucleation load for dislocation burst, occuring at the same level of the indenter displacement

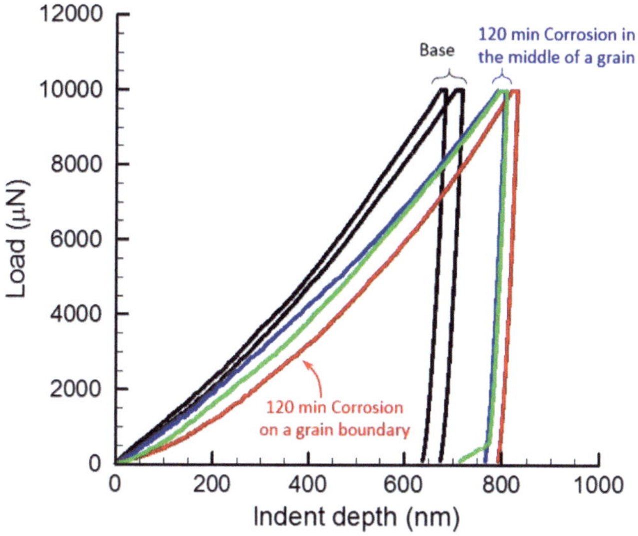

Fig. 5.6 Force-indentation depth curves for the base (no corrosion, *black*) and the 120 min corroded samples (*colored*) after removing the chemical products. A consistent softer response can be seen for the corroded surfaces

A second set of dynamic nanoindentation experiments was performed on the 120 min exposure time, after the corrosion product layer were removed (Fig. 5.2c). All indentations were performed within the middle of grain. Figure 5.6 shows force-indentation depth curves for the base (black curves) sample, and 120-min corroded sample after the cleaning (color curves). A consistent softer response can be seen for the corroded surface. Though, the initial loading segment was almost the same for the reference and the corroded sample, until reaching an indentation depth of about 400 nm. Then, the slope of the loading curves diverges. The maximum indentation depth is about 100 nm deeper for the corroded surface under the same applied load. The indent on a grain boundary was consistently softer.

Figure 5.7 shows the variation of the continuously measured *E*, and *H* as a function of indent depth. Within the experimental error, there is noticeable difference in the measurements of the modulus for the corroded and the base material. The red curve corresponds to an indentation on a grain boundary and exhibiting an initial compliant response with E ~ 75 % of that of the other experiments on either the center of the corroded grain or the reference material. Though an exact value cannot be directly assessed as the measurements near a grain boundary is the convolution of the softer GB response and its thickness [12–14]. The measured hardness exhibit many unique features. First, the continuous decay of *H* for the base material until reaching a steady state level is a direct manifestation of the size dependent response in this range of measurements. Second, the measured *H* is consistently lower for the corroded surface. Third, in the corroded samples, a

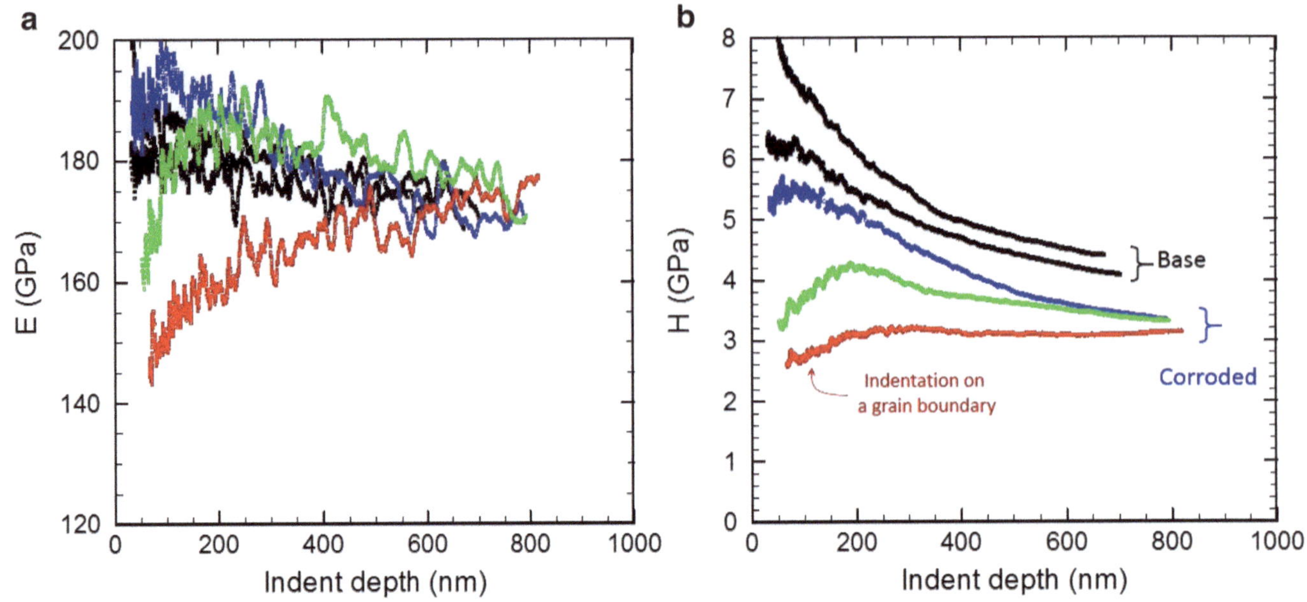

Fig. 5.7 The corresponding indentation (**a**) modulus and (**b**) hardness as a function of indent depth for the base (no corrosion, *black*) and the 120-min corroded samples (*colored*) after 8-min acetic acid treatment of the samples for removal of the corrosion product

transition can be identified in the measured H until reaching a steady state. This transition can be assessed to estimate a thickness of the softer layer near the surface. This layer is estimated to be around ~400 nm. Detailed analysis of the thickness of the soft layer will be performed [15–17].

Figure 5.8 shows two atomic force microscope (AFM) scans of indentations on the pristine reference surface and on the corroded surface at the middle of a grain. It can be seen that the indentation pile-up on the corroded surface is much higher by about 50–80 nm. This is very contrasting with the overall force-indentation curves which were about 100 nm deeper for the corroded case. Such observation can explain the observed similarity in modulus and the great differences in Hardness due to corrosion induced softening. This effect is further explored next.

5.6 Grain Boundary Damage Evolution

The multiple indentation on the center of surface grains highlighted the possibility of softer grain boundaries. To highlight this effect, a set of quasi-static shallow indentation were performed on a line traversing the grain boundary as shown in Fig. 5.9. The indents were performed under a constant load of 1 mN and 1 μm apart. This range of loading would limit the indentation to the initial sink-in range [12, 18] as well as limiting the indentation process zone to about 500 nm. Figure 5.9a is an AFM surface topography rendering of the grain boundaries and the imposed line of the indents. Figure 5.9b is the indentation imprints, shown in the surface gradient mode. The variations of E and H across the GB are shown in Fig. 5.9c, d, respectively. There is an obvious drop in both quantities across the GB from those in the middle of the grain. While the drop is about 15–30 %, however again the actual GB E and H could be much lower as the measurements is greatly affected by the GB thickness [12, 13]. For a typical GB in the range of 0.1–0.3 μm, the properties might be reduced by an order of magnitude from their bulk of the grain values due to the corrosion process.

5.7 Concluding Remarks

The presented study is a systematic exploration of the microstructural and material property changes accompanying the initiation stage of SCC. The mechanical properties of the corrosion product layer, and the near-surface damage evolution were characterized by quasi-static and dynamic nanoindentation technique. The damage evolution within the grain and on

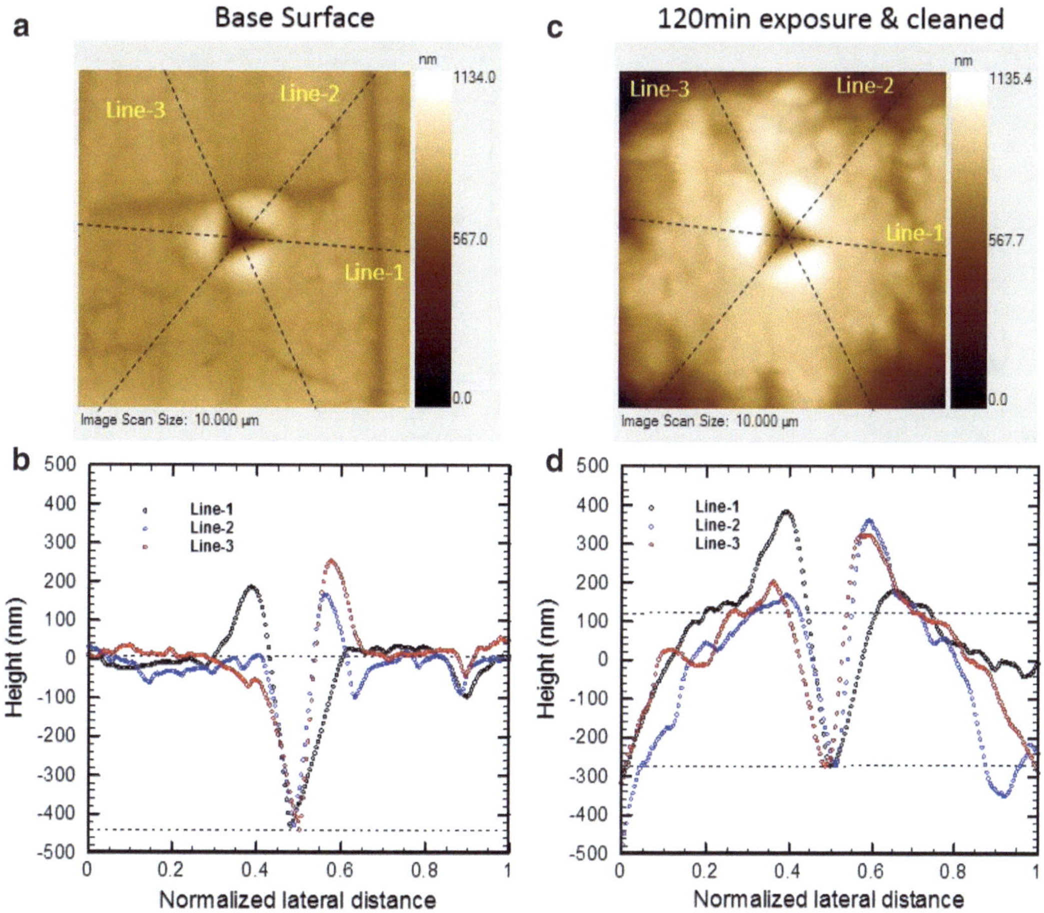

Fig. 5.8 AFM surface scans of the indentation imprints on (a) the base surface with no corrosion, and (b) 120-min corroded sample surfaces. The corresponding height profiles of each line are presented in (c, d), respectively, showing a deeper indent for the base surface

the GBs were shown and the corresponding corrosion mechanisms were discussed. A key finding is the excessive GB weakening with mechanical properties that might be an order of magnitude lower than those of the bulk of the grains due to corrosion. Such excessive weakening of the GBs might be the main driving mechanisms for intergranular SCC in this pH range.

Acknowledgment This work is supported by U.S. Department of Transportation, Pipeline and Hazardous Materials Safety Administration under Competitive Academic Agreement Program No. DTPH5614HCAP03.

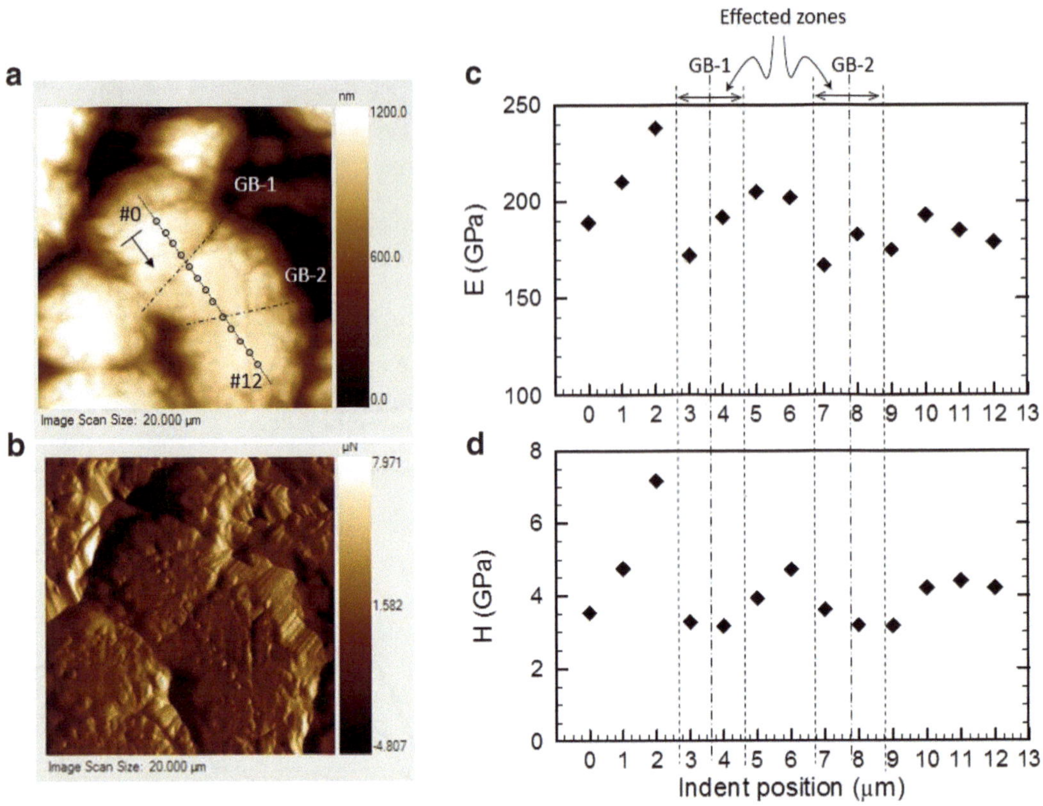

Fig. 5.9 (**a**) AFM surface topology profile of the corroded and cleaned surface, showing the line of indentation, and GBs. (**b**) AFM surface gradient showing the line of indent imprints. (**c, d**) The variation of *E* and *H* across the GB, and showing the relative reduction compared to the bulk of the grain

References

1. Lydell, B.: The Probability of Pipe Failure on the Basis of Operating Experience, vol. 7. ASME, New York (2007)
2. Sieradzki, K., Newman, R.C.: Brittle behavior of ductile metals during stress-corrosion cracking. Philos. Mag. A **51**, 95 (1985)
3. Jones, D.A.: A unified mechanism of stress corrosion and corrosion fatigue cracking. Metall. Trans. A **16**, 1133 (1985)
4. Magnin, T., Chambreuil, A., Bayle, B.: The corrosion-enhanced plasticity model for stress corrosion cracking in ductile FCC alloys. Acta Mater. **44**, 1457 (1996)
5. Parkins, R.N.: Mechanistic aspect of intergranular stress corrosion cracking of high pressure gas pipeline. In: Corrosion 2000, Paper 00363 (2000)
6. Liu, Z.Y., Li, X.G., Du, C.W., Lu, L., Zhang, Y.R., Cheng, Y.F.: Effect of inclusions on initiation of stress corrosion cracks in X70 pipeline steel in an acidic soil environment. Corros. Sci. **51**, 895–900 (2009)
7. Staehle, R.W.: Critical analysis of "tight cracks". Corros. Rev. **28**, 1–103 (2010)
8. Cheng, Y.F., Fu, A.Q.: Electrochemical polarization behavior of X70 steel in thin carbonate/bicarbonate solution layers trapped under a disbonded coating and its implication on pipeline SCC. Corros. Sci. **52**, 2511–2518 (2010)
9. Burstein, G.T., Davis, D.H.: The effects of bicarbonate on the corrosion and passivation of iron. Corrosion **36**, 416–422 (1980)
10. Oliver, W.C., Pharr, G.M.: An improved technique for determining hardness and elastic modulus using load and displacement sensing indentation experiments. J. Mater. Res. **7**, 1564–1583 (1992)
11. Johnson, K.L.: Contact mechanics. Cambridge University Press, Cambridge (2003)
12. Yang, C., Lo, C.T., Narasimhan, B., Bastawros, A.-F.: Measurements of diffusion thickness at polymer interfaces by nanoindentation: a numerically calibrated experimental approach. J. Mater. Res. **24**(3), 970–977 (2009)
13. Bastawros, A., Che, W., Chandra, A.: Measurement of Ultrathin Film Mechanical Properties by Integrated Nano-scratch/indentation Approach. MRS Proceedings 1049, 1049-AA04-04 (2007)
14. Antoniou, A., Bastawros, A.F.: Deformation characteristics of tin-based solder joints. J. Mater. Res. **18**(10), 2304–2309 (2003)
15. Mencik, J., Munz, D., Quandt, E., Weppelmann, E.R., Swain, M.V.: Determination of elastic modulus of thin layers using nanoindentation. J. Mater. Res. **12**(9), 2475–2484 (1997)
16. Saha, R., Nix, W.D.: Effects of the substrate on the determination of thin film mechanical properties by nanoindentation. Acta Mater. 50, 23–38 (1996)
17. Bastawros, A.F.: Analysis of deformation-induced crack tip toughening in ductile single crystals by nano-indentation. Int. J. aol. Struct. **43**(24), 7358–7370 (2006)
18. Taljat, B., Zacharia, T., Pharr, G.M.: Pile-up behavior of spherical indentations in engineering materials. In: Moody, N.R., Gerberich, W.W., Burnham, N., Baker, S.P.: (eds.) Fundamentals of Nanoindentation and Nanotribology, p. 33. (Mater. Res. Soc. Symp. Proc. 522, Warrendale, (1998))

Chapter 6
In-Situ Tensile Test on 316H Sent Using Digital Image Correlation

L. Podesta, B. Wattrisse, F. Latourte, L. Waltz, and J.-M. Muracciole

Abstract At elevated temperature (550 °C) intergranular creep cracks are prone to develop in thermally and environmentally aged 316 stainless steel. To improve the understanding of mechanisms responsible of creep cracking, some micromechanical experiments have been conducted. Single Edge Notched Tensile specimens (SENT) made of 316H have been machined with a desired ratio of $a/W = 0.15$ with a the crack length and W the width specimen. After a fine polishing preparation, the samples are then thermally aged using an oxidizing treatment at 600 °C in a rich-carbon environment during 2000 h. It has been assumed that it induces carbide precipitation at grain boundaries ($Cr_{23}C_6$) leading to a loss of corrosion resistance and to a material embrittlement. In parallel, the oxide (Fe_3O_4) layer grows at the surface up to 50 μm. Its random structure is particularly convenient as a surface marker for the Digital Image Correlation (DIC). A tensile device placed in the SEM chamber is used to apply cycles of loading/unloading to the specimen combining this with a Scanning Electron Microscope (SEM) images acquisitions of the crack tip vicinity. Complementary finite element (FE) simulations of intergranular cracks in bicrystals have been performed and used as reference fields to develop an identification procedure of the crack tip position. It relies on kinematic measurements using a local approach and on projections using Linear Elastic Fracture Mechanic (LEFM) expressions. Experimental evidences have been obtained of plasticity developing ahead of the crack when a local load is applied. A quantification of biais due to the model errors when plasticity is taken into account is done as well as an assessment of the robustness of the procedure. We propose a comparison between FE modeling and full-fields measurements resulting from an *in-situ* tensile test in a pre-cracked sample.

Keywords Austenitic steel • Crystal plasticity • Digital image correlation • Linear elastic fracture mechanic • SEM

6.1 Introduction

At high temperature one of the most detrimental damage mechanism during creep is intergranular cracking. In addition, in harsh environment (CO_2), oxygen and carbon elements diffusion along the grain boundaries is observed. To better understand the crack initiation and propagation in the aged 316H stainless steel, some experiments on microsized specimens are proposed. The creep crack monitoring during *in-situ* tests will be performed combining DIC using high resolution images acquired by microscopy and LEFM expressions. It has been shown that the Williams' series [1] and the Westergaard's solutions [2] are quite efficient to determine the stress intensity factors (SIF) [3, 4] and the crack tip position [5]. Several behaviours have been considered in the current study: elasticity, macroscopic isotropic plasticity and crystal plasticity. Here, the results related to phenomenological crystal plasticity are solely presented. The law chosen to describe the plastic response of the 316H at the grain scale was initially proposed by Méric and Cailletaud [6] and is now commonly used for cyclic loadings in austenitic stainless steels [7]. Crystal plasticity is also relevant to model crystal creep deformation, for example on Ni-based single crystals [8] or in polycrystalline Ti-based alloys such as Ta_6V [9]. In this paper, an identification

L. Podesta (✉)
EDF R&D, Site des Renardières, Avenue des Renardières—Moret-sur-Loing, Ecuelles 77818, France

ThM2, LMGC/CNRS/University Montpellier, 163 Rue Auguste Broussonnet, Montpellier 34090, France
e-mail: laurie.podesta@edf.fr

B. Wattrisse • L. Waltz • J.-M. Muracciole
ThM2, LMGC/CNRS/University Montpellier, 163 Rue Auguste Broussonnet, Montpellier 34090, France

F. Latourte
EDF R&D, Site des Renardières, Avenue des Renardières—Moret-sur-Loing, Ecuelles 77818, France

Y. Zhu, A.T. Zehnder (eds.), *Experimental and Applied Mechanics, Volume 4*,
Conference Proceedings of the Society for Experimental Mechanics Series, DOI 10.1007/978-3-319-42028-8_6

procedure of the crack tip position based on kinematic measurements is presented. It is then validated on real images deformed by a known displacement field associated with the development of crack either for a linear or non-linear behavior, allowing a metrological assessment of the procedure.

6.2 Microsample Preparation

The Grade 316H stainless steel material considered for testing has been in service for 98,704 h at a typical temperature of 500–520 °C. It was not exposed to carbon-rich environment. Therefore the service exposure will only have led to thermal ageing as the service stresses were too low for any significant creep deformation or damage [10]. SENT samples have been machined from this ex-service material with a desired ratio of $a/W = 0.15$ with a the crack length and W the width specimen using Electric Discharge Machining (EDM). Prior to SENT machining, CT samples are submitted to fatigue cycles to propagate a sharp crack from the notch (Fig. 6.1). Then, the CT is cut into 1 mm thick slices using EDM and polished up to 4000 SiC paper to allow the crack tip detection from SEM imaging. The fatigue crack path seems to be inter or transgranular (Fig. 6.2a).

A specific oxidizing treatment, named preconditioning, has been developed to increase the diffusion of oxygen and carbon (C, O) in a reasonable time. After a fine polishing preparation (Fig. 6.2a), the samples are then submitted to preconditioning at 600 °C in CO_2 environment during 2000 h (Fig. 6.2b). In parallel, the oxide (Fe_3O_4) layer growths at the surface up to 50 μm hiding the crack that propagated during the prior fatigue load.

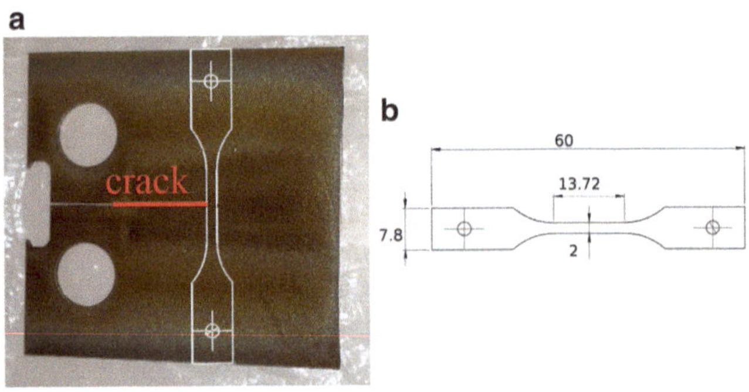

Fig. 6.1 Preliminary machining of pre-cracked CT sample (**a**) before performing the final EDM of the SENT (**b**)

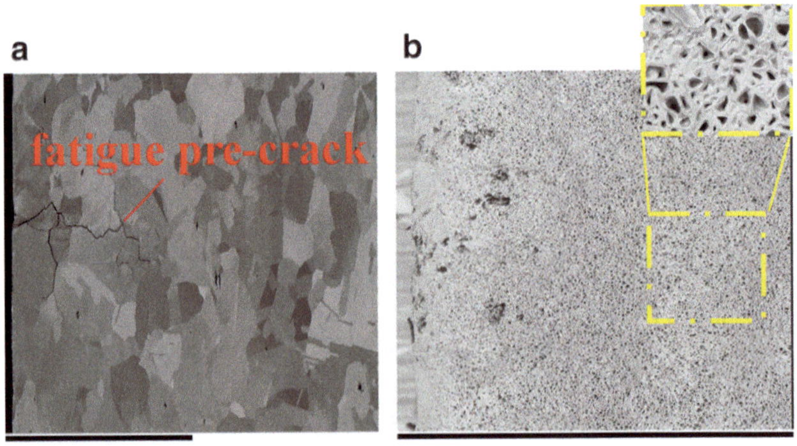

Fig. 6.2 Inter/trans-granular fatigue crack path appears at SENT surface after fine polishing (**a**) but cannot be monitored after preconditioning treatment (**b**). The scale bar is 500 μm and the square length is 35 μm

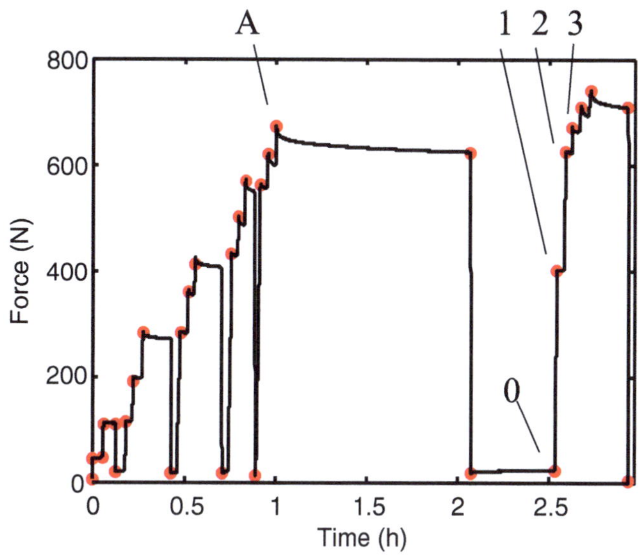

Fig. 6.3 Loading/unloading cycles during the *in-situ* tensile test: loading step A corresponds to the first propagation of the crack, the loading step 0 is defined as a reference state with a load of 20 N allowing to image the crack tip, loading step 1 at 400 N, loading step 2 at 620 N and loading step 3 at 670 N

6.3 Experimental Set-Up

To monitor the crack propagation, a FEI Quanta FEG 600 SEM is used, in combination with a tensile device to apply cycles of loading/unloading. SEM images acquisitions of the crack tip vicinity are conducted. The test is performed at room temperature and vacuum. While the crack propagates, the plasticity development is determined (first step of plasticity, plastic increment, etc...). The loading level is increased as sample surface evolves. Up to loading step A (see Fig. 6.3), no change is visible but a relaxation phenomenon appears, highlighting a viscous behavior. Then, the crack propagation is observed between loading steps 0 and 3. The force acquisition versus time is shown in Fig. 6.3. Red dots indicate image acquisition.

6.4 Kinematic Measurements Using Digital Image Correlation

Kinematic measurements are performed by Digital Image Correlation (DIC) on the surface of a polycrystal of 316H austenitic steel. DIC requires a gray level texture with a dynamic range as large as possible with local contrast variations. In the present case, the random structure of the oxide layer is particularly convenient as a surface marker. The Region Of Interest (ROI) covers a 500 μm side square surface that corresponds to 2048 pixels. Using the same apparatus, the imaging distorsions have been quantified by Guery [11] and are neglected in this work. The displacement fields are measured between two consecutive images considering a local DIC approach on a dense grid. The size of the correlation subsets was 12 pixels and the grid step was 1 pixel. Consequently, the measured displacements are not statistically independent between two adjacent points. Two mechanical quantities of interest are assessed: the oxide deformation prior to crack propagation and the crack advance considering the experimental conditions. They are determined by fitting the displacement field using a specific set of functions related to linear elastic fracture mechanics (*e.g.* Williams' series).

The chosen shape functions for the local DIC algorithm are constant so that the displacement fields obtained by DIC will not be strongly discontinuous due to the finite length scales introduces in the DIC. In other words, the displacement discontinuity due to the crack will be poorly captured.

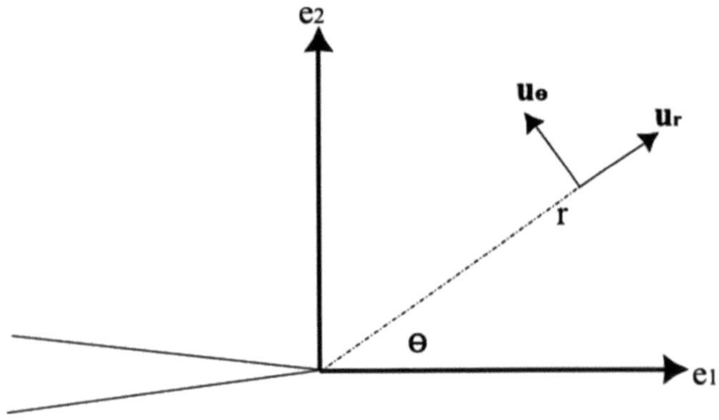

Fig. 6.4 Local coordinate system around the crack tip

Table 6.1 Denomination attributed to each Williams' series orders

Order	n < 0	n = 0	n = 1	n = 2
Designation and mechanical use	Supersingular	Rigid body translation	SIF	Uniform T-stress and both rotation

6.5 Projection Onto Linear Elastic Fracture Mechanic Basis

Williams' series or Westergaard's solutions have been used in Fracture Mechanics to describe the kinematic response of semi-infinite structure under uniaxial loading. Williams' series are used in this work for post-processing the displacement measured by DIC. According to the LEFM theory, the displacement can be expressed, in the polar coordinates, as a combination of contribution from different orders in mode I (opening mode) or mode II (sliding mode) leading to the general equation (Eq. (6.1)). An illustration of the chosen coordinates around the crack tip is proposed in Fig. 6.4.

Using these coordinates, the displacement field corresponding to the Williams' series reads:

$$u(r,\theta) = \sum_{i=I,II} \sum_{n} a_i^n \phi_i^n(r,\theta) \tag{6.1}$$

where n corresponds to the considered order, a_i^n are the target parameters and ϕ_i^n are the reference fields from LEFM, and i is the crack mode. The ϕ_i^n expressions must be separated in agreement with the considered mode: mode I (Eq. (6.2)), mode II (Eq. (6.3)).

$$\phi_I^n = r^{n/2}.\left(\kappa.e^{(jn\theta/2)} - \left(\frac{n}{2}\right).e^{(j(4-n).\theta/2)} + \left(\left(\frac{n}{2}\right) + (-1)^n\right).e^{(-jn\theta/2)}\right) \tag{6.2}$$

$$\phi_{II}^n = jr^{n/2}.\left(\kappa.e^{(jn\theta/2)} + \left(\frac{n}{2}\right).e^{(j(4-n).\theta/2)} - \left(\left(\frac{n}{2}\right) - (-1)^n\right).e^{(-jn\theta/2)}\right) \tag{6.3}$$

where k is the Kolossov's constant ($k = (3 - \upsilon)/(1 + \upsilon)$ in the stress plane conditions, with υ the Poisson coefficient). The different orders are precised in Table 6.1.

In this work, we have chosen to project the displacement fields measured by DIC on a specific set of Williams' series restricted to a mode I crack, using orders 0, 1 and 2.

6.6 Simulation Using A Crystal Plasticity Model

The simulation of the experimental tensile test is performed using the finite element software Code_Aster. The phenomenological crystal plasticity law proposed by Méric and Cailletaud [6] has been chosen in this study. Intergranular cracked bicrystal has been modeling and meshed using quadratic quadrangles (Fig. 6.5a) using the Salome software. To provide numerical convergence, the crack is modelled as a notch of very small radius. A mesh refinement in the notch tip region has been performed (Fig. 6.5b). The crack position is perfectly known. Two boundaries conditions can be used to load the

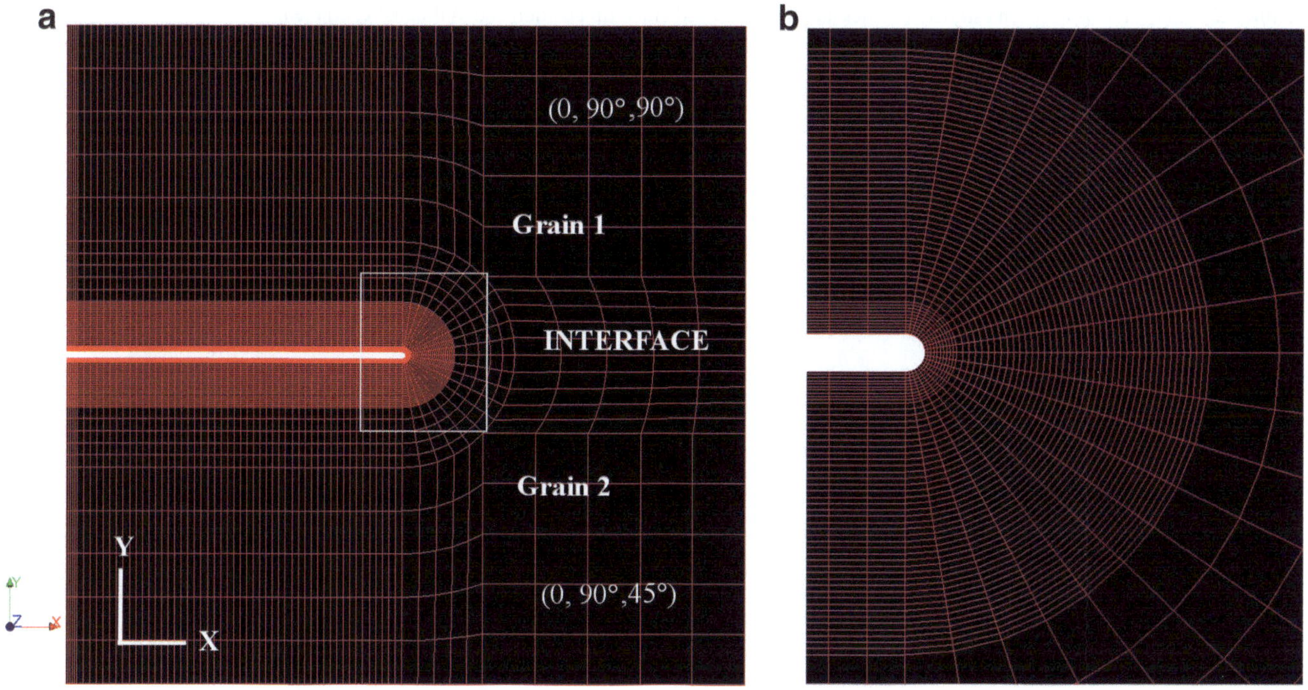

Fig. 6.5 Cracked bicrystal mesh (7850 faces and 23951 nodes) (**a**) corresponding to the cracked bicrystal mesh and the (**b**) corresponding to the mesh refinement performed in the crack tip region

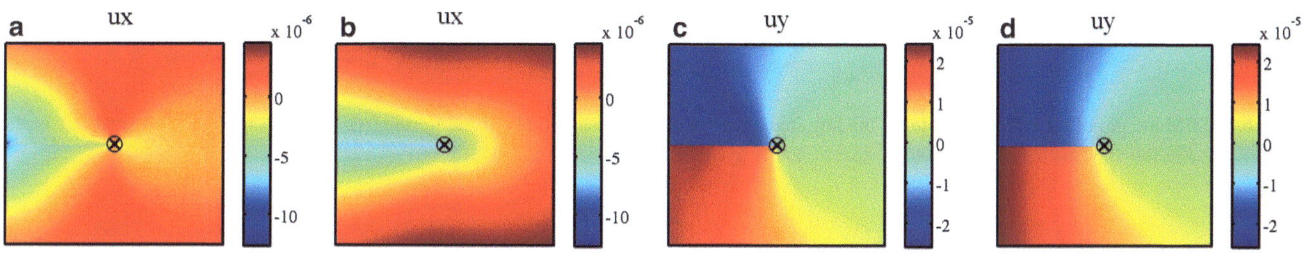

Fig. 6.6 Comparisons between the kinematic fields computed by FE analysis corresponding to Westergaard's boundary conditions (component u_x (**a**) and u_y (**c**)) and the optimized displacement fields from Williams' series provided by Eqs. (6.1), (6.2) and (6.3) (component u_x (**b**) and u_y (**d**))

structure: Westergaard's solutions in displacement prescribed for the global boundary, or imposed displacement on the upper and lower boundaries while the lateral ones are stress-free. It is important to note that this latest case (material properties and boundary conditions) are not consistent with the analytic solutions given by the Williams's series. The crystal plasticity response considering Westergaard's boundaries conditions is solely presented here. Because of the non-dimensional nature of the problem, the width and height of the simulation box are taken equal to 1.

Considering a Face Centered Cubic (FCC) crystals, numerous misorientations have been studied but only one bicrystal example will be presented. The material parameters correspond to a 316L polycrystalline aggregate [7]. The displacement fields corresponding to this numerical study are shown in Fig. 6.6. The fields computed by Code Aster are shown in Fig. 6.6a, c. From these reference solutions, a projection was done using the chosen Williams' series, for an imposed position of the crack tip. The projected fields are shown in Fig. 6.6b, d. The mismatch between the reference fields and the projected ones is attributed to plasticity. Finally, a complete identification of crack parameters (crack position and angle and Williams' series coefficients) has been conducted for different initial guesses of crack tip position, and the results are provided in Table 6.2.

The adimensional positions appearing in Table 6.2 are defined as:

$$\hat{x} = \frac{x}{L_x} \tag{6.4}$$

Table 6.2 Comparisons between estimated values and the optimized crack tip position and Williams' coefficient

	$\hat{x}$	$\hat{y}$	Θ (radian)	$a_1{}^I$ (MPa.√m)
Initial	0	0	0	0.01
Optimized	$-6.9{\cdot}10^{-4}$	$5.1{\cdot}10^{-4}$	$-7.02{\cdot}10^{-5}$	0.015
Initial	0.05	0.05	0.035	0.01
Optimized	$1.6{\cdot}10^{-2}$	$1.2{\cdot}10^{-3}$	$1.00{\cdot}10^{-2}$	0.012

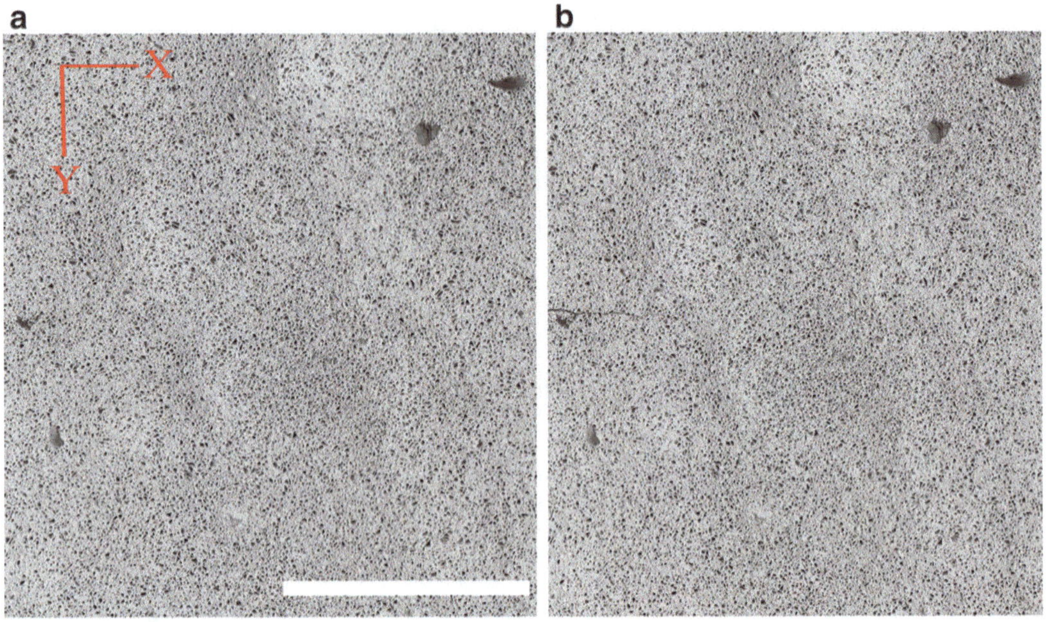

Fig. 6.7 SEM images of the reference state for loading step 0 (**a**) and after propagation for loading step 2 (F = 620 N) (**b**). The physical size of the pixels is 0.244 μm. The scale bar is 250 μm

$$\hat{y} = \frac{y}{L_y} \tag{6.5}$$

with L_x and L_y the investigated area dimensions in the two directions.

6.7 Assessment of the Identification Procedure: The Experimental Test

During the *in-situ* tensile test, the crack propagation in the oxide layer is observed (Fig. 6.7). The Williams' series are not linear with respect to the crack tip position. Consequently, an estimation of the crack tip position has to be introduced in the identification procedure. Here, the crack tip is arbitrarily supposed to be located close to the image center. This estimation can be improved when a crack is visible in the SEM images.

Considering the crack tip position estimation and the Williams' series expressions, a comparison between the DIC results (Fig. 6.8) and the analytical displacement fields can be performed. Then, a least square minimization is used to identify the crack tip parameters and the Williams' series coefficients. The results from the optimization is given in Fig. 6.9. A black circle is used as marker for the optimized crack tip position.

The residuals map shown a good agreement between the reference image and the deformed image corrected by the displacement for the DIC results and the analytical expressions (Fig. 6.10). Table 6.3 summarizes the results of crack parameters identification. The physical size of the pixel is 0.24 μm. The scale bar is 250 μm (1000 pixels).

A quantitative assessment of the results is presented in Table 6.3. The bold values of x_0 highlight the crack propagation as a function of increasing load. For step 2 and 3 at loads of 620 N and 670 N, a same initial guess was chosen for the crack tip position that was not detrimental to capture the crack propagation.

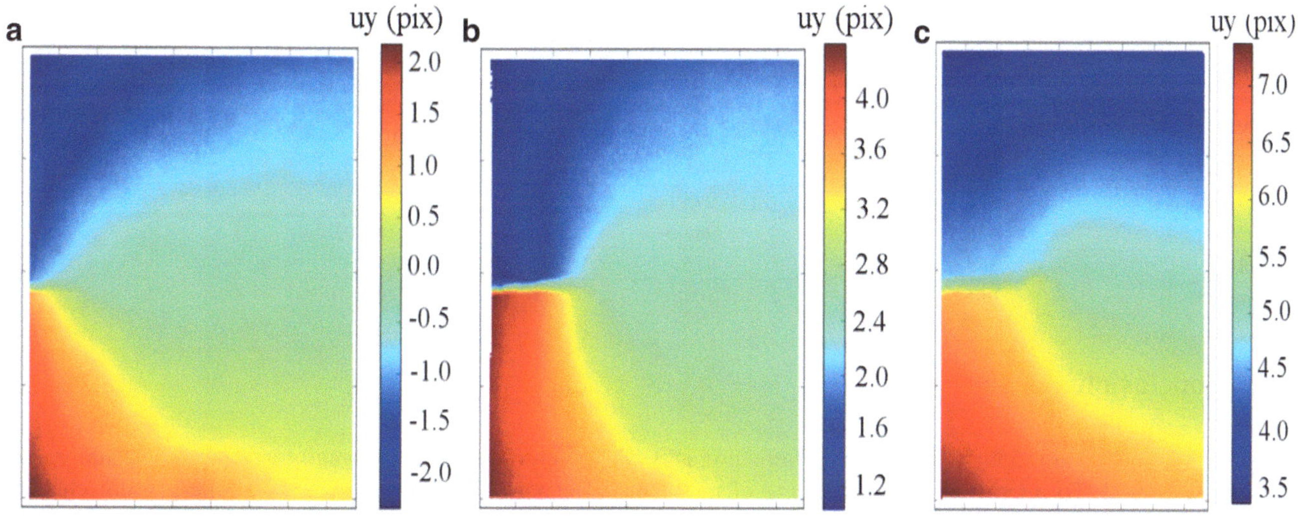

Fig. 6.8 Displacement fields obtained by DIC u_y for loading step 1 at 400 N (**a**), step 2 at 620 N (**b**) and step 3 at 670 N (**c**)

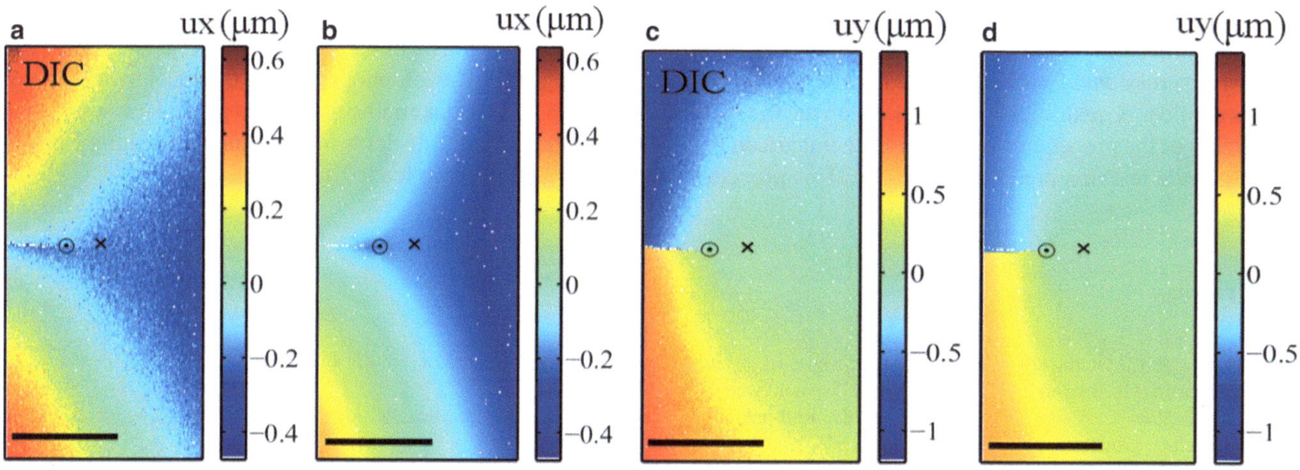

Fig. 6.9 Comparisons between the kinematic fields obtained by DIC (component u_x (**a**) and u_y (**c**)) and the optimized Williams' displacement fields (component u_x (**b**) and u_y (**d**)) for F = 620 N

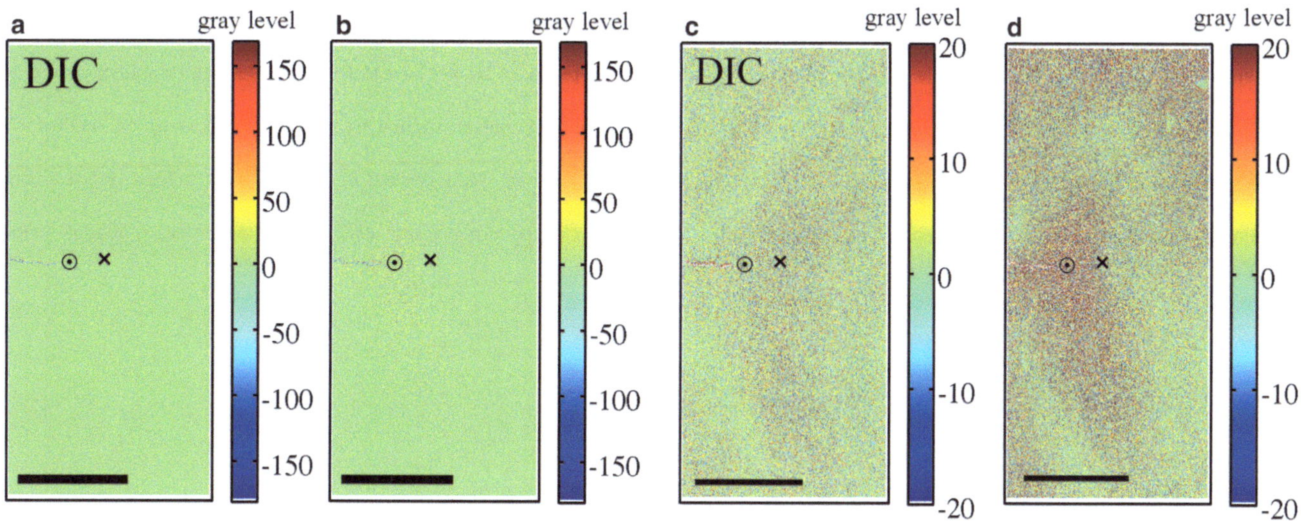

Fig. 6.10 Residual maps for the DIC (**a**, **c**) and the Williams'expressions (**b**, **d**) with a thresholded colorbar of values (**c**, **d**) to highlight the low values of residuals (F = 620 N). The *colorbars* correspond to gray levels coded in 8 bits

Table 6.3 Comparisons between estimated values and the optimized crack tip position

	x_O (pixel)	y_O (pixel)	Θ (radian)	Force (N)
Initial	291	1057	0.00	400
Optimized	**240.1**	981.0	$7.00 \cdot 10^{-5}$	
Initial	835	975	0.00	620
Optimized	**535.6**	986.0	$-2.169 \cdot 10^{-2}$	
Initial	835	975	0.00	670
Optimized	**651.7**	1048.5	$-2.2661 \cdot 10^{-2}$	

6.8 Conclusions

At elevated temperature (550 °C) intergranular creep cracks have been observed in thermally and environmentally aged 316 stainless steel. To improve the understanding of mechanisms responsible of creep cracking, some micromechanical experiments have been conducted. Single Edge Notched Tensile specimens (SENT) made of 316H have been machined with a desired ratio of $a/W = 0.15$ with a the crack length and W the width specimen. A tensile device placed in the SEM chamber is used to apply cycles of loading/unloading to the specimen combining this with a Scanning Electron Microscope (SEM) images acquisitions of the crack tip vicinity. Complementary finite element (FE) simulations of intergranular cracks in bicrystals have been performed and used as reference fields to develop an identification procedure of the crack tip position. It relies on kinematic measurements using a local DIC approach and on projections using Linear Elastic Fracture Mechanic (LEFM) expressions. Experimental evidences have been obtained of plasticity developing ahead of the crack when a local load is applied. A quantification of bias due to the model errors when plasticity is taken into account is possible as well as the robustness of the procedure. We propose a comparison between FE modeling and full-fields measurements resulting from an *in-situ* tensile test in a pre-cracked sample. This procedure will be applied to creep loaded specimens.

References

1. Williams, M.L.: On the stress distribution at the base of a stationary crack. J. Appl. Mech. **24**, 109–114 (1957)
2. Westergaard, H.: Bearing pressure and cracks. J. Appl. Mech. **6**, 49–553 (1939)
3. Roux-Langlois, C., Gravouil, A., Baietto, M.-C., Réthoré, J., Mathieu, F., Hild, F., Roux, S.: DIC identification and X-FEM simulation of fatigue crack growth based on the Williams' series. Int. J. Solids Struct. **53**, 38–47 (2015)
4. Pataky, G.J., Sangid, M.D., Sehitoglu, H., Hamilton, R.F., Maier, H.J., Sofronis, P.: Full field measurements of anisotropic stress intensity factor ranges in fatigue. Eng. Fract. Mech. **94**, 13–28 (2012)
5. Réthoré, J., Roux, S., Hild, F.: Optimal and noise-robust extraction of fracture mechanics parameters from kinematic measurements. Eng. Fract. Mech. **78**, 1827–1845 (2011)
6. Méric, L., Poubanne, P., Cailletaud, G.: Single crystal modeling for structural calculations: part 1—model presentation. J. Eng. Mater. Technol. Trans. ASME **113**, 162–170 (1991)
7. Guilhem, Y., Basseville, S., Curtit, F., Stephan, J.-M., Cailletaud, G.: Numerical investigations of the free surface effect in three dimensional polycrystalline aggregates. Comput. Mater. Sci. **70**, 150–162 (2013)
8. Samal, M.K., Ghosh, S.: Evaluation of creep deformation and mechanical properties of Nickel-based superalloys through FE analysis based on crystal plasticity models. Proc. Eng. **55**, 342–347 (2013)
9. Hasija, V., Ghosh, S., Mills, M.J., Joseph, D.S.: Deformation and creep modeling in polycrystalline Ta_6Al alloys. Acta Mater. **51**, 4533–4549 (2003)
10. Chevalier. M.: Impact of in-service carburization on the creep-fatigue lifetime of stainless steel components and suggested assessment procedure modifications. In: Creep Conference (2015)
11. Guery, A., Latourte, F., Hild, F., Roux, S.: Characterization of SEM speckle pattern marking and imaging distortion by digital image correlation. Meas. Sci. Technol. **25**, 015401 (2014)

Chapter 7
Compression Testing of Silica Microspheres with Synchronized SEM Video

Jennifer Hay

Abstract The Young's moduli of individual silica microspheres, having a diameter of about 4 μm, are measured by compression testing inside an SEM chamber. The testing instrument is a nanoindenter, fitted with a frustum having a face-diameter of 10 μm. While continuously monitoring force and deflection, each particle was compressed to failure. During this semi-static compression, a harmonic force dither having a frequency of 100 Hz was superimposed in order to obtain a continuous measure of particle stiffness. Using this continuous stiffness, the Young's moduli are calculated according to a derivative form of an analytic model for compressing a sphere between two platens (Elastic compression of spheres and cylinders at point contact and line contact, CSIRO National Standards Laboratory Technical Paper, no. 25, 1969). At a compressive strain of 0.2 %, the mean Young's modulus is 73.6 ± 4.4 GPa (N = 12) which is exactly within the range of what we expect for soda-lime glass. The results are enhanced by *in situ* SEM images which reveal both plastic yield and fracture.

Keywords Micro-compression • Nanoindentation • Modulus • Glass • Microsphere

7.1 Introduction

Microspheres have myriad uses in modern life. Hollow microspheres are used to lower the density of manufactured materials. In modern liquid chromatography, the analyte is forced through a column packed with glass microspheres, thus causing components of the analyte to separate based on the speed at which they can pass through the column. In electronic packaging, metalized polymer microspheres are packed together to created flexible, reliable connections. The glass microspheres tested in this work are incorporated into paints to enhance appearance and mar-resistance. In all these applications, knowing the mechanical properties of the microspheres is a critical aspect of design.

With relatively minor changes in hardware and test procedure, the nanoindentation actuator becomes a general small-scale compression testing system. In this case, the actuator was fit with a flat-ended punch in order to set up a test condition that can be analyzed as a double-sided Hertzian contact [1]. For the test geometry illustrated in Fig. 7.1, the Young's modulus, E, is related to the particle diameter, D, compression force, P, and the total compression distance, α, as:

$$E = 3P\left(1 - \nu^2\right)/\left(D^{1/2}\alpha^{3/2}\right). \tag{7.1}$$

Superimposing a small oscillation on the semi-static loading process allows the direct and continuous measurement of the elastic stiffness of the particle, S [2]. This additional information provides an experimental advantage in that one may use a derivative form of Eq. (7.1) ($S = \mathrm{d}P/\mathrm{d}\alpha$) which reveals a more instantaneous relationship between properties and measured parameters:

$$E = 2S\left(1 - \nu^2\right)/\sqrt{\alpha D}. \tag{7.2}$$

In other words, the Young's modulus is related to the instantaneous stiffness, S, and the compression distance, α. Not only is Eq. (7.2) less sensitive to common measurement errors in the absolute values of P and α, it can reveal changes in modulus caused by increasing compressive strain. This is particularly important for polymer spheres, because the compression

J. Hay (✉)
Nanomechanics, Inc., 105 Meco Lane, Oak Ridge, TN 37830, USA
e-mail: jenny.hay@nanomechanicsinc.com

Y. Zhu, A.T. Zehnder (eds.), *Experimental and Applied Mechanics, Volume 4*,
Conference Proceedings of the Society for Experimental Mechanics Series, DOI 10.1007/978-3-319-42028-8_7

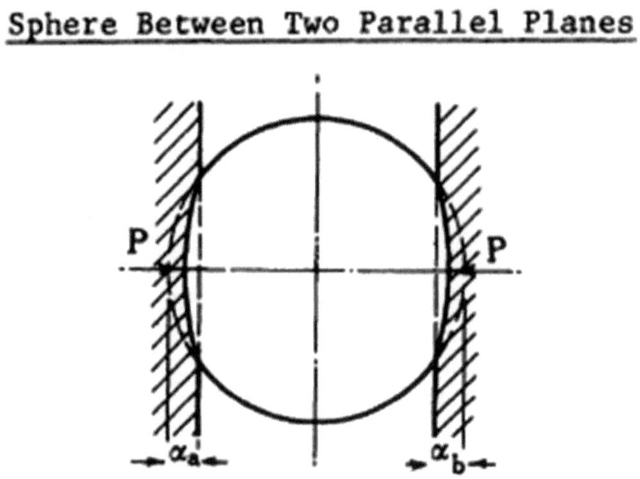

Fig 7.1 Compression of a sphere between two infinitely stiff platens, analyzed as a double-sided Hertzian contact [1]. In this work, the face of the flat punch is one platen and the substrate is the other

process causes the polymer chains to stretch and become more closely packed, which can manifest as a true increase in modulus. In other words, the compressed polymer microsphere generally has a higher modulus than the original microsphere. Thus, Eq. (7.2) is a better model, but it can only be used with testing instruments that superimpose an oscillatory force on the semi-static compression process.

To normalize results for particle size, we employ a simple definition of strain,

$$\epsilon = \frac{\alpha}{D}, \tag{7.3}$$

and report modulus for each microsphere at a specific value of strain.

7.2 Experimental Method

All tests were performed with the NanoFlip produced by Nanomechanics, Inc. (Oak Ridge, TN), which includes an inForce 50 actuator. The NanoFlip was chosen because its small size allows testing within an SEM chamber, and its superior positioning capability allows motion control with nanometer-scale precision. The actuator was configured with a tip extender, and fit with a conductive diamond frustum ($\psi = 60^{\circ}$, $R = 5\mu m$). The entire testing system was placed inside the main chamber of a Philips XL 30 Scanning Electron Microscope as shown schematically in Fig. 7.2 to allow simultaneous imaging and testing.

Solid microspheres made from recycled soda-lime glass (mean diameter of 4 μm) were acquired from Prizmalite Industries, Inc.[1] Microspheres were dispersed on a glass microscope slide using ethanol.

All tests were performed using the test method "InSEM Dynamic Spherical Compression." This test method operates in individual mode, allowing the user to execute one test at a time. The following steps were performed for each test:

1. Guided by a video stream from the SEM, the moving platen (the flat face of the diamond frustum indenter) was centered over the target particle, approximately 2 μm above the top of the particle.
2. The diameter of the particle was measured using a high-quality SEM image.
3. The moving platen approached the particle at a velocity of 50 nm/s until contact was detected.
4. The particle was loaded at a rate of 0.1 mN/s while simultaneously oscillating the moving platen with an amplitude of 2 nm and a frequency of 100 Hz. Loading was terminated when the total compression exceeded 1500 nm or the applied force exceeded 50 mN.
5. The particle was unloaded over the course of 1 s.

[1] Prizmalite Industries, Inc. 380 Lexington Ave, 17th Floor, New York, NYU 10168. Product Name: P2011SL Glass Microspheres; Lot Number: PS130803A.

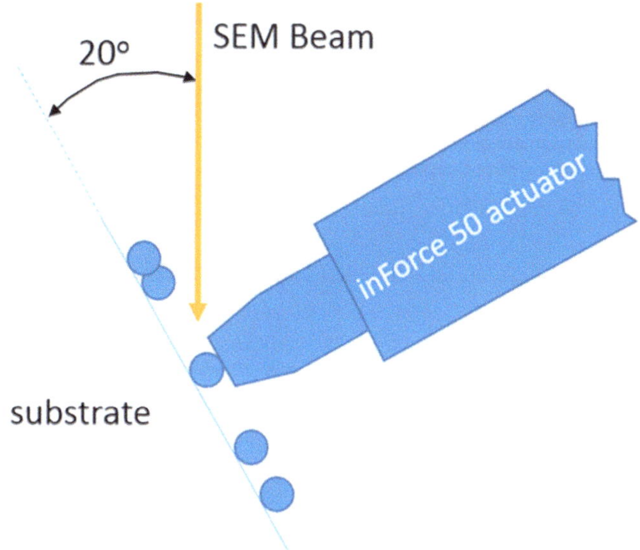

Fig 7.2 Schematic of test set-up

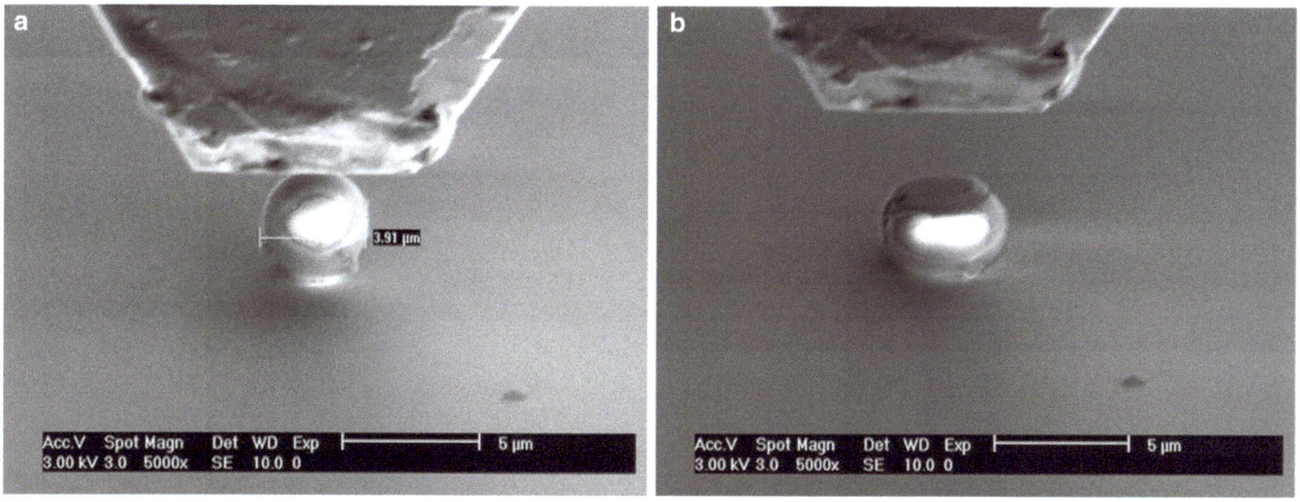

Fig 7.3 (**a**) Microsphere 1 before testing and (**b**) after testing; particle has both yielded plastically and fractured

During the approach, loading, and unloading, fast-scan video from the SEM was automatically acquired and synchronized with the mechanical data in the testing software.

7.3 Results and Discussion

Figure 7.3 shows the SEM images of the first microsphere, before and after testing. (Other microspheres exhibit similar deformation.) Clearly, the test causes both plastic yield, as indicated by the flattened shape, as well as fracture. Although the goal of this work is to measure Young's modulus, future efforts will focus on measuring yield strength by means of a similar test.

The superimposed oscillation we apply during loading (1 nm, 100 Hz) yields a continuous measure of stiffness as each microsphere is compressed. From this continuous measure of stiffness, modulus is calculated according to Eq. (7.2). Figure 7.4 shows this calculation of modulus as a continuous function of strain for all tests. As expected, the measured modulus is more consistent from test to test at larger strains. This is because inaccuracies in the compression distance (which

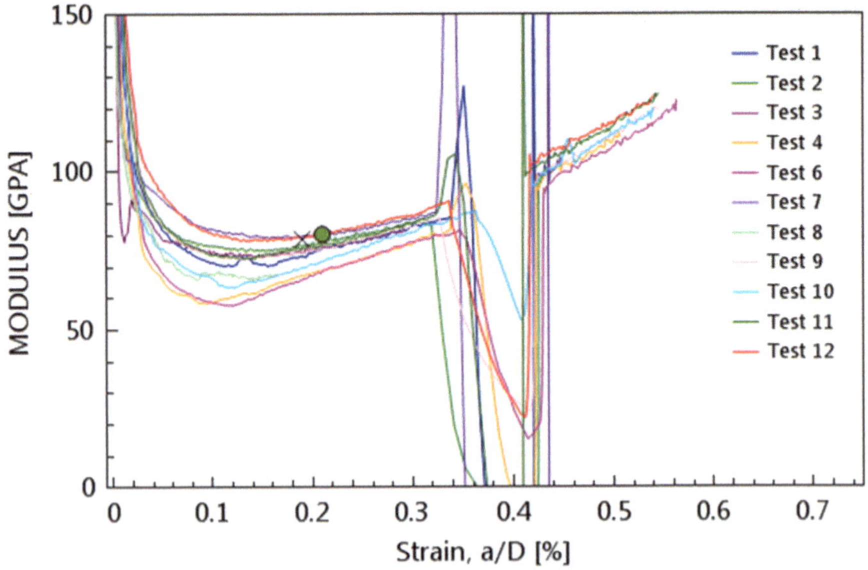

Fig 7.4 Modulus as a function of compressive strain for 12 microspheres. Modulus is reported at 0.2 % compressive strain. Failure of the microsphere is obvious at strains just greater than 0.3 %

Table 7.1 Summary of diameter and Young's modulus for soda-lime glass microspheres

Microsphere	Particle diameter (μm)	E @0.25 % strain (GPa)
1	3.91	74.44
2	4.31	77.16
3	5.44	75.57
4	3.12	68.20
5	4.34	71.36
6	2.79	67.31
7	3.55	80.01
8	7.40	67.29
9	2.89	75.24
10	2.92	70.75
11	2.89	76.38
12	2.92	79.80
Median	**3.34**	**74.84**
Mean	**3.87**	**73.63**
Standard Deviation	**1.32**	**4.32**

arise primarily from imperfections in the shapes of the contacting surfaces) become relatively unimportant as the overall compression distance increases. All particles fail at a strain just above 0.3 %. Thus, we report the measured modulus at a lesser strain around 0.20 % (between the "X" and the green circle in Fig. 7.4).

Figure 7.4 illustrates a very practical advantage of the superimposed oscillation: simplicity in testing. We did not need to know much about the sample or its expected behavior in order to plan a good test. The superimposed oscillation gives properties as a continuous function of strain (or load, or depth) and the experimenter is free to decide, after the fact, the ideal conditions for reporting properties—in this case, just prior to fracture. Nanomechanics' software makes these advantages available to the experimenter by allowing changes in data analysis after testing is complete.

Table 7.1 summarizes the measurements of diameter and modulus for all microspheres. Even for this very small (and not random) sampling, the mean diameter is close to that reported by the manufacturer (4 μm). And the measured modulus of 73.6 GPa is within the range of what we should expect for soda-lime glass (72–74 GPa).

7.4 Conclusions

The NanoFlip nanoindenter from Nanomechanics, Inc. is used as a general micro-scale compression tester in order to measure the Young's modulus of glass microspheres of about 4 μm in diameter. The test is interpreted with a double-sided Hertzian model in order to achieve a value for Young's modulus. The dynamic aspect of this testing system, which measures elastic stiffness as a continuous function of compression, reduces sensitivity to imperfections in the shapes of the contacting bodies, thus allowing an accurate measurement of Young's modulus (73.6 ± 4.4 GPa).

References

1. Puttock, M.J., Thwaite, E.G.: Elastic compression of spheres and cylinders at point contact and line contact. CSIRO National Standards Laboratory Technical Paper, no. 25 (1969)
2. Oliver, W.C., Pharr, G.M.: An improved technique for determining hardness and elastic modulus using load and displacement sensing indentation experiments. J. Mater. Res. **7**(6), 1564–1583 (1992)

Chapter 8
Development of Glass/Steel Bibeam Specimen for Study of Brittle Crack Path Stability

S.J. Grutzik and E.D. Reedy

Abstract We have developed a novel specimen for studying crack paths in glass. Under certain conditions, the specimen reaches a state where the crack must select between multiple paths satisfying the $K_{II} = 0$ condition. This path selection is a simple but difficult benchmark case for both analytical and numerical methods of predicting crack propagation. We document the development of the specimen, using an uncracked and instrumented test case to study the effect of adhesive choice and validate the accuracy of both a simple beam theory model and a finite element model. In addition, we present preliminary fracture test results and provide a comparison to the path predicted by two numerical methods (mesh restructuring and XFEM).

Keywords Brittle fracture • Glass • Stability • Crack path

8.1 Introduction

Predicting the path along which a brittle crack will propagate is an important but difficult problem. A number of criteria have been proposed since the 1960s to predict crack propagation under mixed mode loading which include maximum hoop stress (or maximum tensile stress) [1], minimum strain energy density [2], and maximum energy release rate [3]. All of these criteria predict that a crack under pure mode I loading will propagate straight ahead. Abstract here In other words, a crack will propagate in a direction such that $K_{II} = 0$ where K_{II} is the mode II stress intensity factor. The issue of whether a crack path is stable, if a small deviation from the $K_{II} = 0$ path will bring the crack back in line with the original path or cause further deviation, is a different question. Cotterell and Rice [4] proposed that the $K_{II} = 0$ path is stable when $T < 0$, where T is uniform tensile stress parallel to the crack direction superposed over the $r^{-1/2}$ asymptotic crack tip field [5]. T is in fact the r^0 term of the asymptotic expansion developed Williams [6]. Yang and Ravi-Chandar [7] have shown the $T < 0$ condition to be of limited applicability. We have developed a novel, thermally loaded specimen which achieves a condition where two independent crack growth directions satisfy the $K_{II} = 0$ criterion. These paths intersect at a 90° angle and a crack initiated along one can switch to the other with no nominal preference for the crack to turn to the left or to the right. This specimen allows for easy study of the factors controlling crack path stability and the sensitivity of the crack path to various geometrical and material parameters thus providing simple but difficult benchmark case for numerical methods of predicting crack path such as extended finite elements (XFEM)[8] and adaptive remeshing [9].

8.2 Specimen Preparation

The specimens consist of nominally 300 mm beams of borosilicate glass and 304 stainless steel with nominal cross sections 6 mm × 3.17 mm and 6 mm × 31.7 mm respectively joined with an adhesive. Epon 828/Epikure 3140 (100/45 pbw) and U-05FL Hysol urethane adhesive were chosen based on their extremely different mechanical properties and ability to cure at room temperature to test the effect of adhesive choice on specimen behavior. Monofilament nylon spacers were placed in the bond line at the edges of the bars during epoxy curing to provide a uniform bond thickness of approximately 30 μm. After

S.J. Grutzik (✉)
Materials Mechanics and Tribology, Sandia National Laboratories, Albuquerque NM 87185, USA
e-mail: sjgrutz@sandia.gov

E.D. Reedy
Component Science and Mechanics, Sandia National Laboratories, Albuquerque NM 87185, USA

Y. Zhu, A.T. Zehnder (eds.), *Experimental and Applied Mechanics, Volume 4*,
Conference Proceedings of the Society for Experimental Mechanics Series, DOI 10.1007/978-3-319-42028-8_8

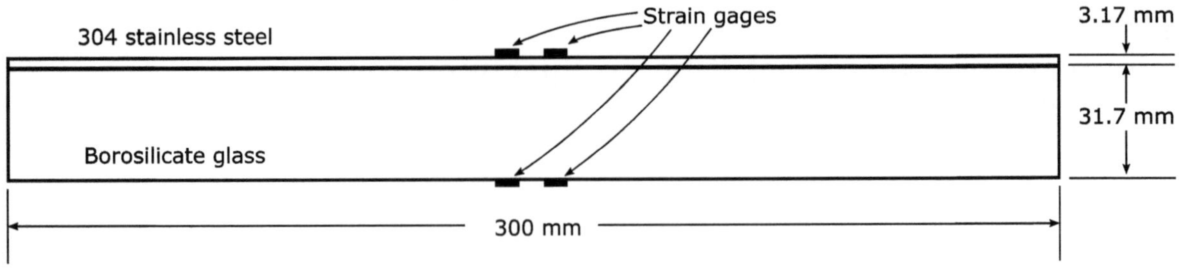

Fig. 8.1 Schematic of the bibeam specimen with strain gage locations indicated

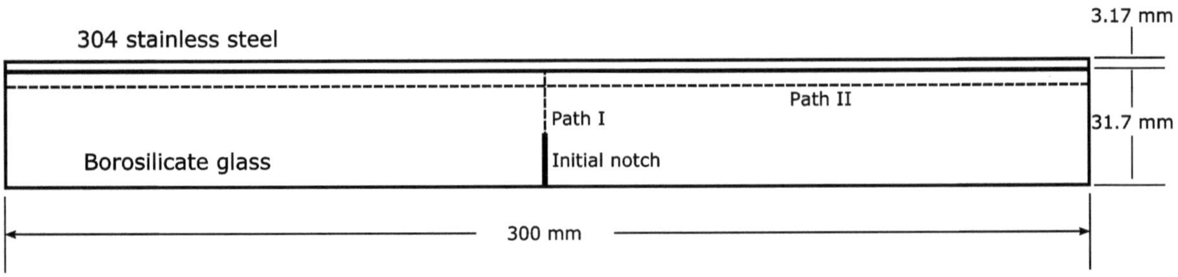

Fig. 8.2 Diagram of the glass/metal bibeam specimen showing the initial notch location and two paths satisfying $K_{II} = 0$

curing, the ends of the bars which included the nylon spacers were cut off to remove any possible stress concentrations in the bond line. To validate the specimen design and ensure negligible impact of the adhesive on the desired stress state in the bi-material beam an uncracked and instrumented specimen with each adhesive was fabricated. Strain gages were placed on the outer edges of both the bonded glass and steel beams as well as equivalent locations on unbounded reference beams of the same material. Temperature was measured with thermocouples placed near the strain gages on each beam. 350 Ω A type strain gauges were used to best match to the thermal expansion of the stainless steel at temperatures between −30 °C and 30 °C and minimize self heating effects. Specimens were either cooled slowly between these maximum and minimum temperature or in steps of 10 °C held for 1 h. A schematic of the specimen is shown in Fig. 8.1. For fracture testing, the initial notch was created using a rotary diamond blade. The specimen was held at 45° relative to the blade and cut from both sides to create a chevron notch. During the initial loading this chevron fills first resulting in a sharp crack front. A schematic of the notched specimen is shown in Fig. 8.2.

8.3 Strain Gage Calibration

Strain gage output is temperature sensitive due to the temperature dependence on the resistivity of the gage materials and thermal mismatch between the gage and the test sample. Gages are provided with a calibration to account for this sensitivity, however the calibration and gage design is optimized for a particular structural material (usually steel) and a range of test temperatures near room temperature [10, 11].

A preliminary check on strain gage accuracy through the temperature range used in these tests can be obtained using the strain gage output on the unbonded reference bars alone. If the output of the gage on the glass reference bar is subtracted from that of the gage on the steel reference bar, the result is the difference in thermal expansion. The slope of the resulting curves can be compared to the difference in CTE of the two materials. This comparison is shown in Fig. 8.3. Two gages placed on both the glass and steel, resulting in measured CTE differences of 12.4 μm/m/°C and 12.6 μm/m/°C. The difference between accepted CTE values is 14.1 μm/m/○C. The CTE difference measured by the two sets of gages was 2.4 %. A comparison of the measured and accepted CTE values for glass and steel yielded an average difference of 11 % between those values. The agreement between the accepted and measured differences was regarded as sufficient, effectively qualifying the strain gages for the crack growth study.

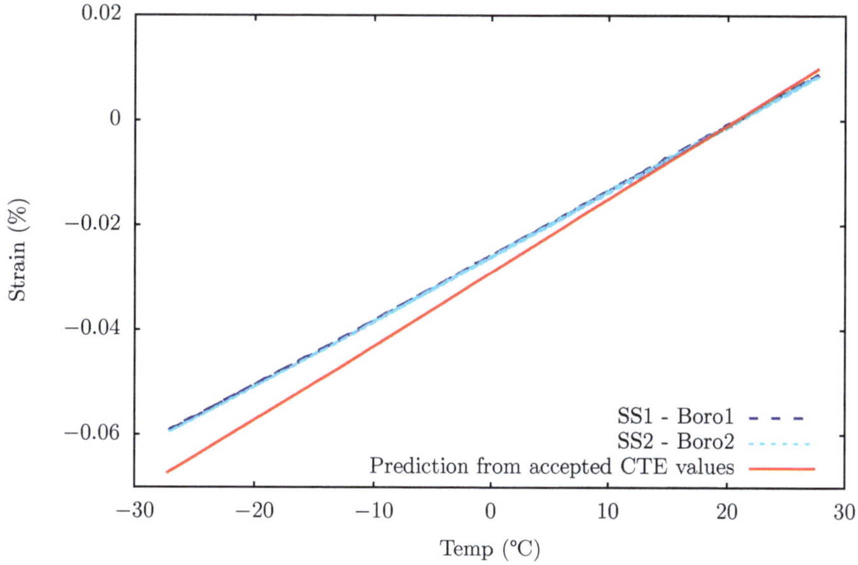

Fig. 8.3 Difference in thermal expansion as measured by strain gages on reference materials as well as the prediction from accepted values

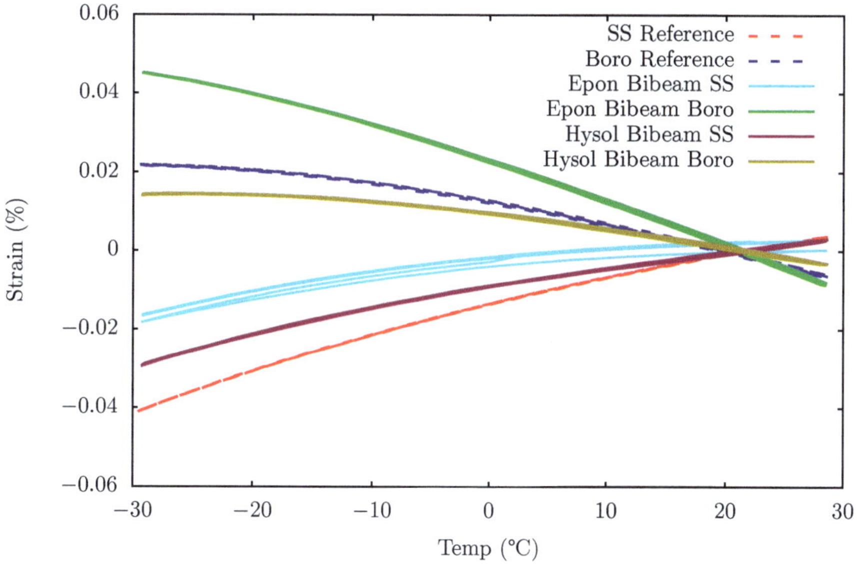

Fig. 8.4 Raw strain gage output for bibeams with Epon epoxy (E Bi) and Hysol urethane adhesive (U Bi) as well as unbonded glass and steel reference bars

8.4 Adhesive Selection

Preliminary characterization through strain gage instrumentation and photoelastic stress measurements indicated that the adhesive could impact the stress distribution in the bi-material beams. A small study was performed to optimize the application of the adhesive such that its impact on the stress distribution of the bi-material beams was minimized. Bibeam specimens were made with two different adhesives; the above mentioned Epon 828/Epikure 3140 epoxy and U-05FL Hysol urethane adhesive. Compared to the Epon/Epikure epoxy, the Hysol urethane adhesive provides a better bond to the glass surface but is significantly more compliant. A bibeam specimen with each adhesive was prepared and instrumented with strain gages and thermocouples as described above and illustrated in Fig. 8.1. The bibeam specimens were simultaneously cooled to −30 °C producing the strain gage results shown in Fig. 8.4, where E Bi is a bibeam with the Epon epoxy and U Bi is a bibeam with the Hysol urethane adhesive. Even though the adhesive layer is thin (about 30 μm as set by the nylon spacers), the response of the bibeam is quite different between the two adhesives. A finite element model was used to further understand the effect of adhesive choice on bibeam performance. Figure 8.5 shows finite element calculated strains at the glass surface with both adhesives modeled as either a linear elastic or Neo-Hookean [12] hyperelastic material. The linear elastic moduli and CTEs used in the finite element model are listed in Table 8.1. The Neo-Hookean parameters for the two adhesives are listed in Table 8.2. The bibeam geometry used in the simulations was slightly different than the bibeam used to

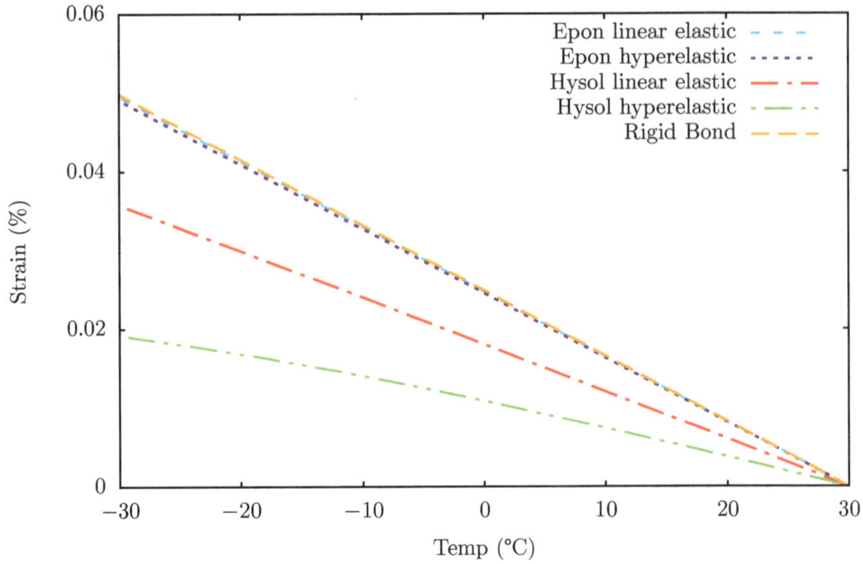

Fig. 8.5 Finite element calculated stress and strain for a bibeam with either Epon epoxy or Hysol urethane adhesive modeled as both linear elastic or Neo-Hookean hyperelastic material as well as assuming a rigid bond at the interface

Table 8.1 Material properties for bibeam materials with U-04FL estimated to be 100 times more compliant than Epon 828

	E (GPa)	ν	CTE (μm/m/°C)
Borosilicate glass [13]	64	0.2	3.25
304 Stainless steel [14]	193	0.29	17.3
Epon 828/Epikure 3140 epoxy [15]	2.9	0.4	NA
U-04FL Hysol urethane adhesive	0.029	0.45	NA

Table 8.2 Neo-Hookean parameters for the two adhesives with Epon 828 C_{10} estimated from a tensile strength of 8.5 ksi and elongation at failure of 4.5 % [15] and D_1 as $10 \cdot C_{10}$ and U-04FL estimated as 100 times more compliant than Epon 828

	C_{10} (MPa)	D_1 (MPa)
Epon 828/Epikure 3140 epoxy	166	1660
U-04FL Hysol urethane adhesive	1.66	16.6

produce the experimental result give in Fig. 8.4. Thus, the magnitude of the strain provided by the simulations was not expected to provide a direct match with that experimental result. From Fig. 8.5, we see that the adhesive choice has a large effect on the response of a bibeam. The results show that not only does a compliant adhesive impact the bibeam response but the constitutive details of the compliance have a magnified effect on bibeam response. Furthermore, it is apparent that the details of the constitutive response of the Epon can be neglected when calculating stresses and strains. For these reasons, it was decided that all future bibeams will be made with the Epon epoxy rather than the Hysol urethane adhesive, even though the bond strength is weaker.

8.5 Validation of Bibeam with Epon Epoxy

If the Epon adhesive layer is assumed to be a rigid bond, the layer can be neglected and a bibeam specimen with no cracks can be modeled using Euler-Bernoulli beam theory. As a final validation, a bibeam specimen was made with the Epon epoxy and two nominally identical strain gages from the same package were placed at each measurement location. If the output of a gage at an equivalent location on an unbonded reference bar is subtracted from the output of a gage on the bibeam, the result is linear in temperature and can be compared with Euler-Bernoulli beam theory prediction. The comparison is shown in Fig. 8.6. The strain as calculated with Euler-Bernoulli beam theory is also shown. There is a 5.1 % difference between the two gages on the steel beam. An error occurred with one of the gages on the glass beam such that no measure of repeatability can be found for gages on the glass. There is an average difference of 14.3 % between the gages on the steel and the beam theory prediction and an error of 5.3 % between the gage on the glass and the beam theory prediction. The larger difference in the beam theory of the gage on the steel can be accounted for by the larger relative uncertainty of the dimensions of the thin steel beam.

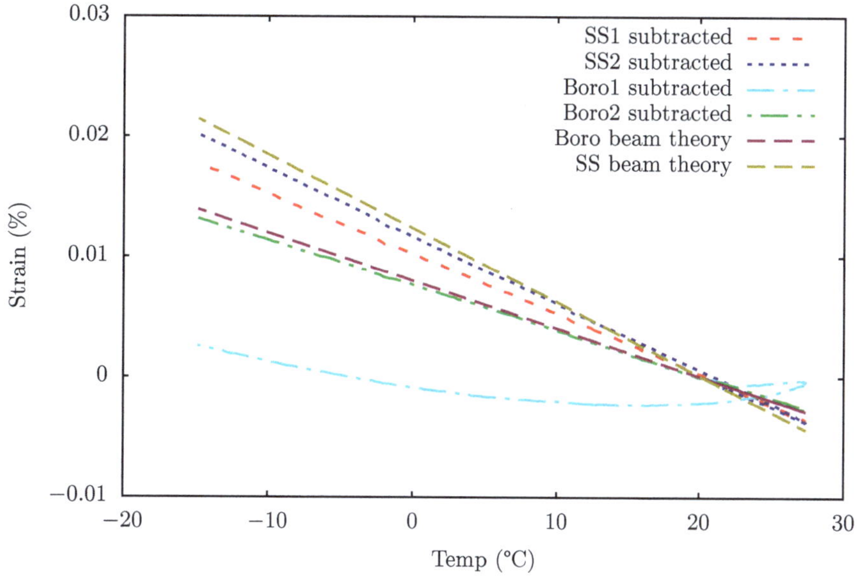

Fig. 8.6 Corrected strain gage output on the glass and steel parts of the bibeam along with beam theory predictions

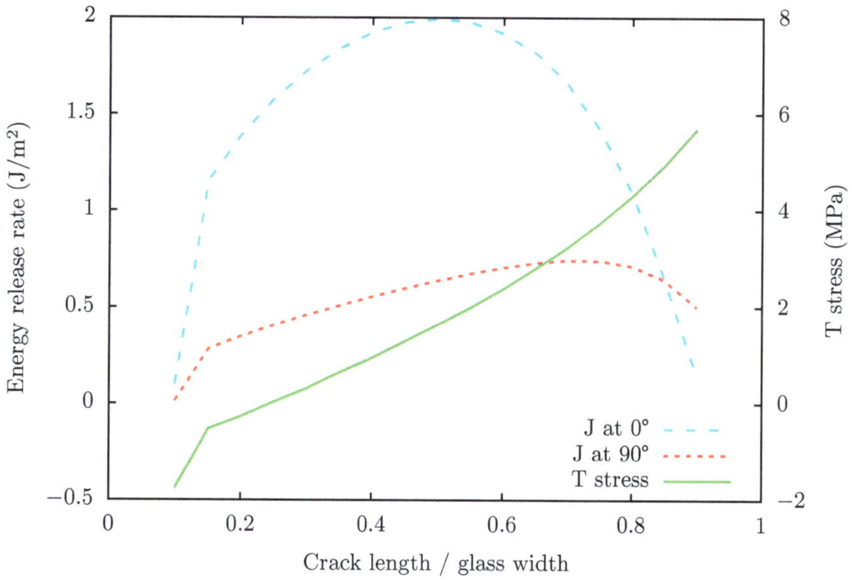

Fig. 8.7 T stress and energy release rate as a function of crack length

8.6 Fracture Testing Procedure

During testing the specimen is cooled from room temperature. The difference in CTE between the steel and the borosilicate puts the specimen in bending. Figure 8.7 shows finite element calculated energy release rate as a function of crack length if the crack is started from the bottom in the middle of the specimen, as depicted in Fig. 8.2. Energy release rate is shown both with a straight crack and a small 90° kink at the crack tip with kink length 1 % of the nominal crack length. If the initial crack is more than roughly half width of the glass, energy release rate for the straight crack decreases with increasing crack length and the crack propagates stably with further loading (decreasing temperature).

Due to the placement of the cooling fan there is a slight temperature gradient in the testing chamber. To counteract this, the specimen is suspended freely inside a secondary chamber made of clear acrylic. The secondary chamber removes any convective cooling effect from the fan and results in a uniform temperature distribution along the length of the specimen. Specimen temperature is tracked with a thermocouple placed on the surface of the glass. The stress field around the crack tip is visualized via the photoelastic effect [16]. This gives a precise location of the crack tip and eventually will allow stress intensity factors and T-stress to be extracted directly from the photoelastic image. The arrangement of the photoelastic polariscope is depicted in Fig. 8.8.

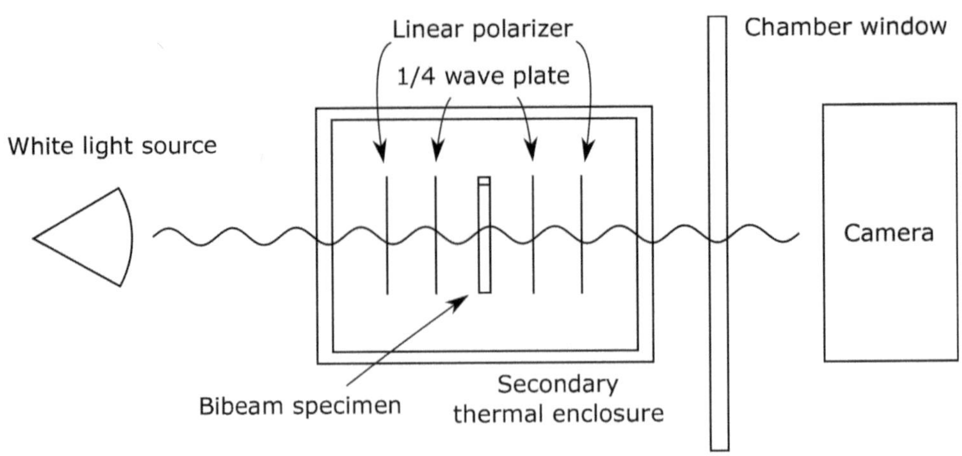

Fig. 8.8 Diagram showing the arrangement of the polariscope used for crack path observation

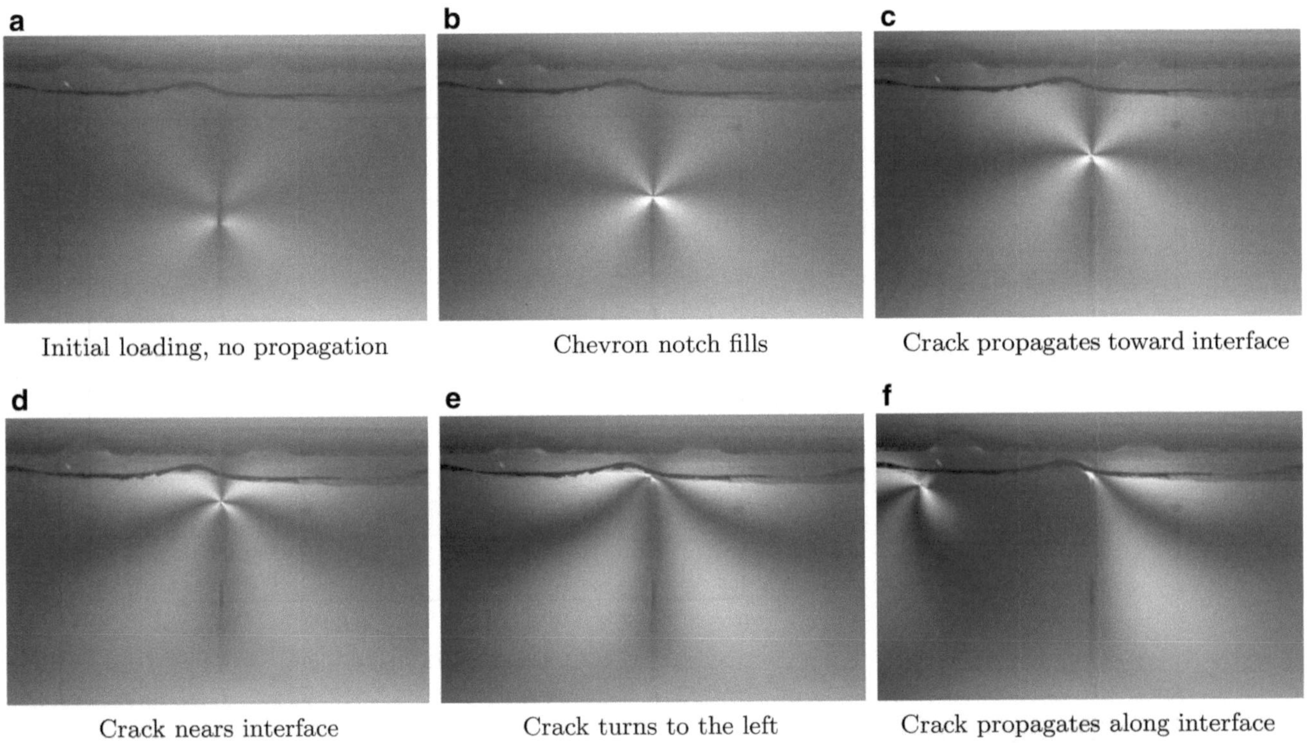

Fig. 8.9 An example of a crack propagation sequence with the crack tip highlighted by photoelasticity. (**a**) Initial loading, no propagation. (**b**) Chevron notch fills. (**c**) Crack propagates toward interface. (**d**) Crack nears interface. (**e**) Crack turns to the left. (**f**) Crack propagates along interface

8.7 Fracture Testing Results

A series of photoelastic images as the crack propagates during cooling is shown in Fig. 8.9. A numerical method to track the crack tip in each image frame and extract the stress intensity factors and T-stress [17] is currently in development. As the crack tip approaches the steel, the crack path becomes unstable and the crack growth direction switches to the alternate $K_{II} = 0$ path parallel to the interface (as depicted in Fig. 8.2). As the crack tip approaches the steel the initial crack path becomes unstable (although it still satisfies $K_{II} = 0$) and the crack switches to the alternate path parallel to the interface. The distance between this second crack path and the interface matches that predicted for steady state cracking in a brittle

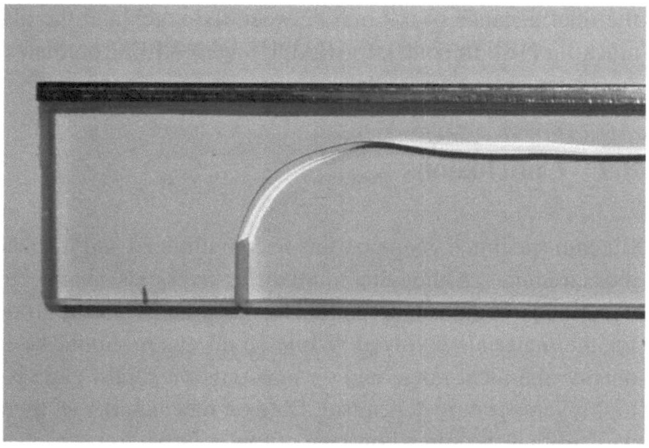

Fig. 8.10 Crack propagating from a notch located near the end of a bibeam specimen

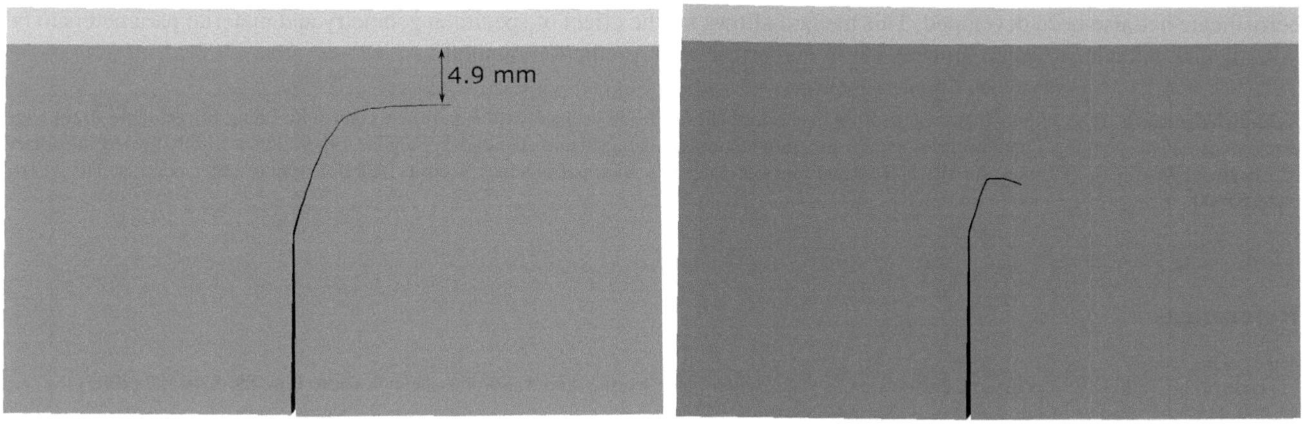

Fig. 8.11 Simulation of crack propagation in an end-notched specimen with SIERRA/SM with FRANC3D (*left*) and ABAQUS with XFEM (*right*)

substrate by Suo and Hutchinson [18]. If the crack is initiated at the center of the specimen the stress field is nominally symmetric about the crack tip and there is no preference to it turning left or right.

The cause of this behavior is suggested by the energy release rate curves in Fig. 8.7. Along with energy release rate for a straight crack, Fig. 8.7 also shows energy release rate with a small 90° kink at the end of the crack and *T*-stress. When the crack nears the glass/steel interface, energy release rate of the kinked crack (parallel to the interface) becomes greater than the energy release rate of the straight crack. This suggests the straight crack is no longer stable beyond this point and any small deviation from a straight path will cause it to continue to turn until the crack is propagating parallel to the interface. *T*-stress is positive during the entire process, except for very small initial notches. The Cotterell and Rice $T > 0$ criterion for crack path instability is thus satisfied at all times and is overly conservative in this case since path instability only occurs near the interface. If the crack is initiated near the end of the specimen, the stress field is not symmetric and the crack always turns toward the interior of the specimen at a larger radius. This is shown in Fig. 8.10.

8.8 Comparison with Numerical Crack Propagation Calculations

Some initial attempts have been made to numerically predict the crack path in the asymmetric case where the crack is initiated near the end of the specimen. The specimen shown in Fig. 8.10 where the notch is located near the end of the specimen was modeled using the XFEM capabilities of the commercial FEA code ABAQUS [19] and Sandia's SIERRA/SolidMechanics [20] used in conjunction with the commercial fracture code FRANC3D [21]. XFEM models cracks without mesh restructuring, calculating the crack faces within each element with nodal shape functions. FRANC3D [21] uses adaptive mesh restructuring to incrementally advance a crack through a model. The crack path predicted by each method is shown in Fig. 8.11. The SIERRA/FRANC3D result not only qualitatively matches the experiment (shown in Fig. 8.10) but the final distance of 4.9 mm between the crack and the glass/metal interface agrees with an analytical model of substrate cracking [18]. In contrast, ABAQUS with XFEM predicts the crack turning parallel to the interface much too early.

8.9 Conclusions

Bibeam specimen geometry has been optimized and its behavior validated for fracture-based experiments using strain gage measurements. Although a small differences exists between the strain gage results and beam theory they can be attributed to uncertainty in thermal response and repeatability of the strain gages and spread in reported values for elastic moduli and CTE for the materials involved. While strain gages should be fairly repeatable under the right circumstances, at temperatures outside the ideal range and on materials the strain gage is not compensated for, the thermal behavior is complicated and highly temperature dependant. Data on repeatability of thermal effects in strain gages is sparse to nonexistent. Nonetheless, we feel that the spread of gage output relative to each other and predictions from accepted material parameters and beam theory are reasonable given the factors involved. A experimental method providing stable crack propagation through the borosilicate has also been developed. This method allows for the effect of specimen geometry and material parameters to be investigated and comparisons can be made to numerical crack prediction methods.

Acknowledgements The authors thank Garth Rohr and Corey Gibson for assistance with experiments and John Laing for valuable strain gage discussions. Sandia National Laboratories is a multi-program laboratory managed and operated by Sandia Corporation, a wholly owned subsidiary of Lockheed Martin Corporation, for the U.S. Department of Energy's National Nuclear Security Administration under contract DE-AC04-04AL85000.

References

1. Erdogan, F., Sih, G.C.: On the crack extension in plates under plane loading and transverse shear. J. Basic Eng. **85**, 516–527 (1963)
2. Sih, G.C.: Mechanics of Fracture, vol. 1. Springer, Berlin (1972)
3. Wu, C.W.: Maximum energy release rate criterion applied to a tension-compression specimen with crack. J. Elast. **8** (3), 235–257 (1978)
4. Cotterell, B., Rice, J.R.: Slightly curved or kinked cracks. Int. J. Fract. **16** (2), 155–169 (1980)
5. Gupta, M., Alderliesten, R.C., Benedictus, R.: A review of T-stress and its effects in fracture mechanics. Eng. Fract. Mech. **134**, 218–241 (2015)
6. Williams, M.L.: On stress distribution at base of stationary crack. J. Appl. Mech. **24** (56), 109–114 (1957)
7. Yang, B., Ravi-Chandar, K.: Crack path instabilities in a quenched glass plate. J. Mech. Phys. Solids **49** (1), 91–130 (2001)
8. Moës, N., Dolbow, J., Belytschko, T.: A finite element method for crack growth without remeshing. Int. J. Numer. Methods Eng. **46** (1), 131–150 (1999)
9. Bittencourt, T.N., Wawrzynek, P.A., Ingraffea, A.R., Sousa, J.L.: Quasi-automatic simulation of crack propagation for 2D LEFM problems. Eng. Fract. Mech. **55** (2), 321–334 (1996)
10. Micro-Measurements: Measurement of thermal expansion coefficient using strain gages. Vishay Precision Group, Technical Report TN-513-1 (2010)
11. Micro-Measurements: Strain gage thermal output and gage factor variation with temperature. Vishay Precision Group, Technical Report TN-504-1 (2014)
12. ABAQUS Analysis User's Guide, 6th edn., sect. 22.5.1. SIMULIA (2014)
13. Corning, Inc: Properties of PYREX®, PRYEXPLUS®, and low actinic PYREX code 7740 glasses
14. Committee, A.I.H.: Metals Handbook, vol. 1, 10th edn. ASM World Headquarters, 9639 Kinsman Road Materials Park, OH 44073-0002 (1990)
15. Hexion: Technical data sheet, Epikure™curing agent 3140 (2007). www.hexion.com/Products/TechnicalDataSheet.aspx?id=2654
16. Ramesh, K.: Digital Photoelasticitiy. Springer, Berlin (2000)
17. Singh, V.K., Gope, P.C.: Experimental evaluation of mixed mode stress intensity factor for prediction of crack growth by photoelastic method. J. Fail Anal. Prev. **13**, 217–226 (2013)
18. Suo, Z., Hutchinson, J.W.: Steady-state cracking in brittle substrates beneath adherent films. Int. J. Solids Struct. **25** (11), 1337–1353 (1989)
19. ABAQUS Analyis User's Guide, 6th edn. SIMULIA (2014)
20. Sierra/SolidMechanics User's Guide, 4th edn. Sandia National Laboratories (2015)
21. FRANC3D Reference Manual, 7th edn. Fracture Analysis Consultants, Inc (2016)

Chapter 9
Construction Procedure of Spot Weld Failure Model for Crash Simulation

J. Ha, J.-H. Lim, and C.-Y. Oh

Abstract This paper proposed the construction methodology of spot weld failure model for a crash simulation based on the experiments and the simulations which is called a hybrid method. The test procedure was designed to obtain the failure load of a spot weld under combined loading conditions. The failure surface is determined based on bending moment, normal force and shear force. Those components acting on the spot welded part are obtained from finite element analysis results, which are defined by the real experimental conditions. The proposed failure surface was constructed based on Wung [Wung, Exp Mech 41, 107–113, 2001; Wung et al., Exp Mech 41, 100–106, 2001] model except torsion term. It was found that the failure surface of mild steel was expressed as a function of previous researches, however failure surface of high strength steels and advanced high strength steels were different shape from previous research results. The proposed failure criterion is well-estimated for specimen levels such as cross tension, loading angle of 30°, loading angle of 45°, coach-peel and lap-shear tests.

Keywords Spot weld • Failure model • Hybrid method • Prediction equation • Crash simulation

9.1 Introduction

Improvement of crashworthiness becomes one of the challenging issues in auto industries together with the light weight design. Actual crash test cost to evaluate the crashworthiness of the auto-body structure become larger due to tighten regulation for a car crash. Thus, computer simulation as an alternative method to evaluate the crashworthiness of the auto-body structure is widely used in the automotive industry [3]. In order to estimate the crashworthiness properly, the correct failure prediction of a spot weld is indispensable for the crash simulation [4, 5]. Rupture of a spot weld is likely to occur prior to failure of the base metal when a large load is applied to the auto-body structure since extremely high stress is concentrated at the interface between the nugget and the base metal [6]. Because the impact load transferred from one part to another part through a spot weld is abruptly changed after the spot weld fails, deformation behaviors of the auto-body structure usually reveal large discrepancies between experiment and simulation after joined components are separated.

Research on the failure characteristics of a spot weld has been investigated over the past few decades. Especially, it is necessary to estimate the strength of spot welds under various loading modes in order to provide a failure criterion of a spot weld for the structural analysis or crashworthiness assessment of the auto-body structures using the finite element analysis. Lee et al. [7], Barkey and Kang [8], Madasamy et al. [9] and Langrand and Combescure [10] proposed testing fixtures to provide various loading conditions including pure axial, mixed axial/shear, or pure shear loads on a spot-welded specimen through changing the position of the fixture. The failure strengths of the spot weld under combined loading conditions were utilized to provide the failure criterion. The coefficients that constitute a force-based failure criterion were determined by a regression analysis from the failure strength data of the spot weld. Similarly, Lin et al. [11, 12] proposed a test methodology with a different type of test fixture. They analyzed and clarified the failure mechanism of spot welds in square-cup specimens made from the mild steel and the HSLA steel under combined loading conditions. After failure tests of the spot-welded specimens using four fixture sets designed by them, an engineering failure criterion was proposed to be in quadratic form in terms of the normalized axial and shear loads with consideration of the sheet thickness and the nugget radius under combined

J. Ha (✉) • J.-H. Lim
POSCO, Steel Business Division Steel Solution Marketing Department Materials Forming Research Group, 100, Songdogwahak-ro, Yeonsu-gu, Incheon 21985, South Korea
e-mail: hhuh@kaist.ac.kr; jiwoong99@gmail.com; hajiwoong@posco.com

C.-Y. Oh
KIMS, Industrial Technology Support and Safety Division, Joining Technology Department, 797, Changwon-daero, Seongsan-gu, Changwon-si, Gyeongsangnam-do 51508, South Korea

Y. Zhu, A.T. Zehnder (eds.), *Experimental and Applied Mechanics, Volume 4*,
Conference Proceedings of the Society for Experimental Mechanics Series, DOI 10.1007/978-3-319-42028-8_9

loading conditions. Although the testing fixtures proposed by previous researchers could impose various combinations of axial and shear loads on the spot weld efficiently by changing the loading position of the fixtures, it is impossible to consider the change of load histories acting on the spot weld due to the rotation of the nugget during the failure tests. Since the rotation of the nugget during the failure tests alters the ratio of the axial load and the shear load acting on the spot weld, the change of the load histories by the rotation of the nugget have to be considered when the ration of the axial load and the shear load is calculated using the failure load.

Recently, spot weld failure models are proposed based on the stress acting on the spot welded parts and microscopic approaches. Seeger et al. [13, 14] proposed a stress-based failure criterion which was determined by a regression analysis from the failure test results. In order to construct the failure surface, KSII tests from 0° to 90° and peel test were conducted. Yang et al. [15] proposed a testing protocol for the purpose of creating spot weld failure parameters of a stress-based failure criterion which is implemented in LS-DYNA software. Sandahl et al. [16] proposed the semi-physical modeling method which can predict the spot weld failure and macro cracks.

This paper proposes the construction procedure for the purpose of determining spot weld failure parameters suggested in this research for automotive manufacturers. Failure simulations in the specimen models show that the failure parameters created by this procedure are accurate for predicting the initiation of spot weld failure by comparing results with experimental results.

9.2 Developed Spot Weld Failure Model

Chao [17] proposed a failure criterion based on the failure loads of cross-tension and lap-shear specimens. It is, however, insufficient to provide an accurate failure criterion that describes the behavior of spot welds under combined loading conditions, because spot welds in the auto-body structures are subjected to a complicated loading condition with deformation by car crash. Therefore, a large researcher proposed testing fixtures to provide various loading conditions including pure normal, mixed normal/shear, or pure-shear loads on a spot-welded specimen by changing the position of the fixture.

The coefficients that constitute a force-based failure criterion were determined by a regression analysis from the failure strength data of the spot weld. Lee et al. [7] proposed a test methodology under the combined loading conditions and the spot weld failure model based on experimental results. The failure criterion is expressed as

$$\left(\frac{f_s}{F_S}\right)^n + \left(\frac{f_n}{F_N}\right)^n = 1 \tag{9.1}$$

Here, F_N and F_S are the normal failure load and the shear failure load of a spot weld, respectively. The variable n is a shape parameter. The coefficients that constitute their failure model are obtained using the least square method to minimize the discrepancy between the experimental data and interpolated data. Figure 9.1 shows the proposed spot weld failure model.

In the result of Lee et al. [7], spot weld failure criterions are composed of the normal failure load and the shear failure load. Wung [1] and Wung et al. [2], however, suggested the failure mechanism based on the normal load, shear load, bending and torsion. Wung [1] defined the failure modes of a spot weld by three kinds of mechanism, and proposed the failure criterion based on a failure force. The failure criterion is expressed as

$$\left(\frac{f_s}{F_S}\right)^\alpha + \left(\frac{m_b}{M_b}\right)^\gamma + \left(\frac{f_n}{F_N}\right)^\mu + \left(\frac{m_t}{M_t}\right)^\beta = 1 \tag{9.2}$$

Here, F_N, F_S, M_b and M_t are the normal failure load, the shear failure load, the failure moment and the failure torsion of a spot weld, respectively. The variables of α, β, γ and μ are shape parameters. The coefficients that constitute their failure model are obtained using the least square method to minimize the discrepancy between the experimental data and interpolated data.

The coefficients that constitute a stress-based failure criterion were determined by a regression analysis from the failure strength data of the spot weld. This 3-D failure criterion describes a polynomial failure surface which is proposed by Seeger et al. [13]. The spot weld fails if the stress triple of the internal normal, bending and shear stresses is above the surface. The failure criterion is expressed as

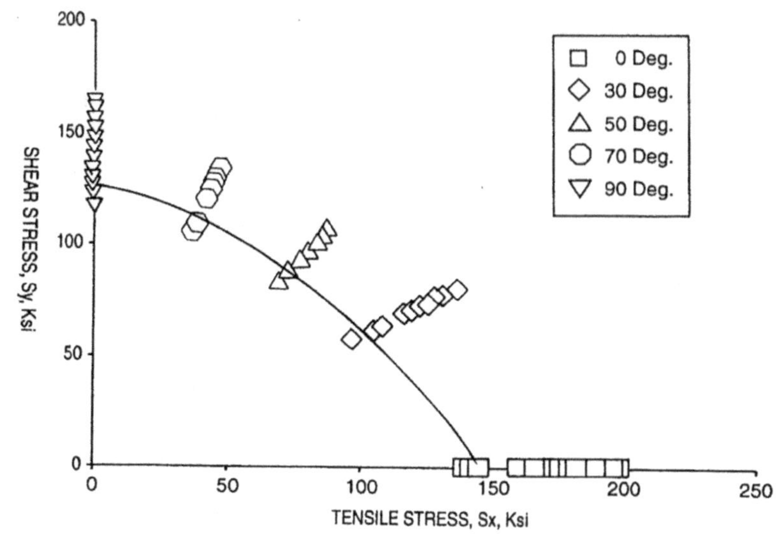

Fig. 9.1 Spot weld failure model proposed by Lee et al. [9]

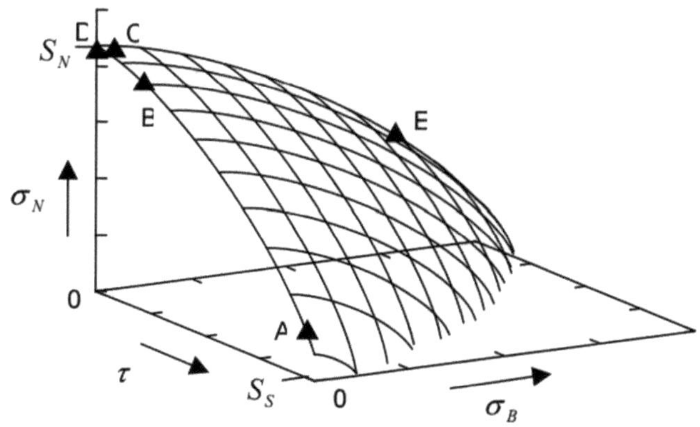

Fig. 9.2 Spot weld failure model proposed by Seeger et al. [13, 14]

$$\left(\frac{\sigma_N}{S_N}\right)^{n_N} + \left(\frac{\sigma_B}{S_B}\right)^{n_B} + \left(\frac{\tau}{S_S}\right)^{n_S} = 1 \tag{9.3}$$

Here, σ_N, σ_B, τ are the normal failure stress, the bending failure stress, the shear failure stress of a spot weld, respectively. The variables of n_N, n_B and n_S are shape parameters. The coefficients that constitute their failure model are obtained using the least square method to minimize the discrepancy between the experimental data and interpolated data.

Dynamic effects is included by strain rate depended strength components as shown in Eq. (9.4).

$$\left(\frac{\sigma_N(\dot{\varepsilon})}{S_N(\dot{\varepsilon})}\right)^{n_N} + \left(\frac{\sigma_B(\dot{\varepsilon})}{S_B(\dot{\varepsilon})}\right)^{n_B} + \left(\frac{\tau(\dot{\varepsilon})}{S_S(\dot{\varepsilon})}\right)^{n_S} = 1 \tag{9.4}$$

Here, $\dot{\varepsilon}$ is effective strain rate, and $S_N(\dot{\varepsilon})$, $S_B(\dot{\varepsilon})$ and $S_S(\dot{\varepsilon})$ are strain-rate dependent normal, bending and shear strength. The strain rate functions are defined by load curves. Figure 9.2 shows the stress-based failure model.

By providing a set of failure parameters for each failure model, failure surfaces can be defined by Eqs. (9.1)–(9.4). If a force state or a stress state are outside the surface, the spot weld will fail in a crash simulation. These failure parameters are specified by each weld size and material stack-up.

9.3 Experimental Procedure for Newly Proposed Spot Weld Failure Model

Prior to spot welding of a specimen, the specimen was wiped with dilute acetone solution using a cloth in order to remove grease and dirt from its surface. Spot welding was then performed using a static spot/projection welding machine. The welding conditions were determined after several U-tension tests with the industry standards to guarantee a button-type failure. In this paper, it is assumed that influential factors on spot weld failure are normal load, shear load and bending, because torsion can be negligible in the automotive structure. Based on this assumption, the spot weld failure model is expressed as

$$\left(\frac{f_n}{F_N}\right)^{\alpha}+\left(\frac{f_s}{F_S}\right)^{\beta}+\left(\frac{m_b}{M_b}\right)^{\gamma}=1 \tag{9.5}$$

Here, F_N, F_S and M_b are the normal failure load, the shear failure load and the failure moment of a spot weld, respectively.

The variables of α, β and γ are shape parameters. In order to obtain the normal, shear and bending failure loads, failure tests of the spot welds were conducted at different initial loading angles of 0°, 30°, 45°, 60° and lap-shear test was done using the testing fixture and specimens. Pure-shear test at a loading angle of 90° was carried out using the fixture and specimen proposed by Ha and Huh [18]. The loading angle indicates the imposed angle of a spot-welded specimen with respect to the loading direction. In addition, lap-shear tests were conducted in order to obtain the failure loads of spot welds. Testing procedures are as shown in Fig. 9.3. Failure tests were conducted using an INSTRON 5583 device with a cross-head speed of 3.0 mm/min until the specimen was separated into two components. The load and the displacement were measured simultaneously at each test. The load was measured with the load cell equipped in the testing machine and the displacement was calculated from the relative movement of the two pull bars.

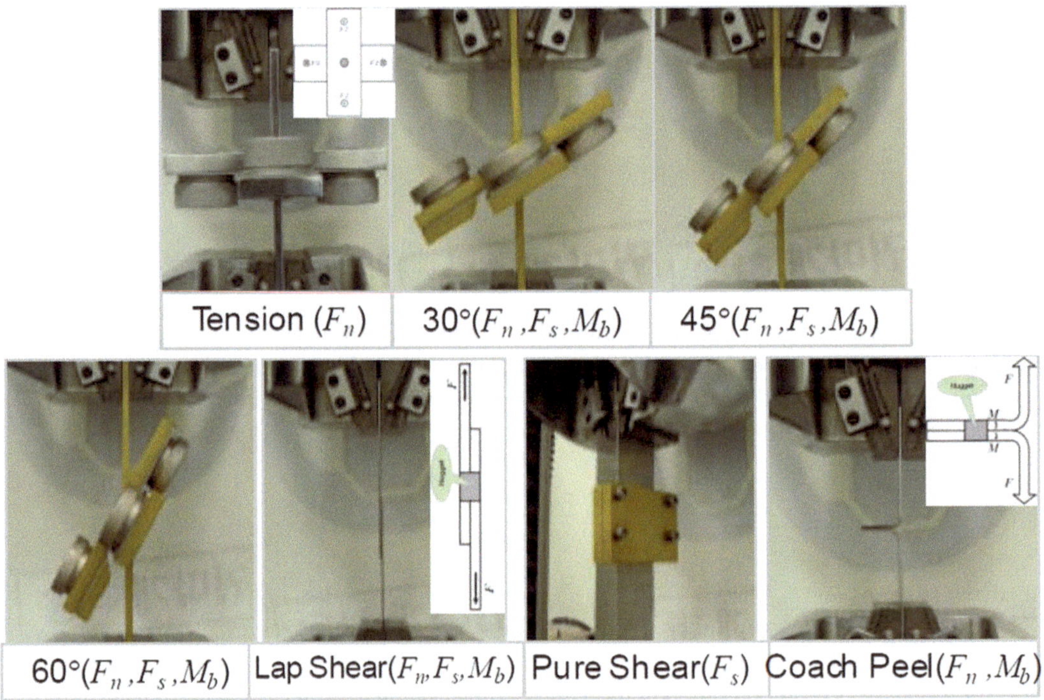

Fig. 9.3 Failure test procedure for newly proposed failure model

9.4 Hybrid Method to Determine the Coefficients of Newly Proposed Failure Model

It is impossible to determine the coefficients of proposed failure model directly from the experiments because combined loads acts on spot welds during failure tests. Acting loads on spot weld are shown in Fig. 9.3 with respect to the testing conditions. In order to determine the failure loads and shape parameters of failure model, decomposing failure loads have to be conducted by the hybrid experimental-numerical procedure which is called as the hybrid method [19].

Hybrid method is utilized to determine the onset of fracture of specimen. The displacement fields on the specimen surface are measured using either two- or three-dimensional digital image correlation (DIC). Based on the DIC measurements, the instant of onset of fracture (not the location) is defined by the first detectable discontinuity in the measured displacement field at the specimen surface. Subsequently, a finite element simulation is performed for each experiment. Post-processing of those simulations gives then access to the evolution of the stress triaxiality and the equivalent plastic strain.

Hybrid method based on the failure loads is utilized in order to obtain the failure load components with respect to failure test conditions. Based on the failure loads obtained in failure tests, the instant of onset of spot weld failure is determined. Subsequently, a finite element simulation is performed for each experiment. Post-processing of those simulations gives failure load components acting on spot welds such as normal, shear and bending loads. These failure load components are plotted on the plane consisting of normal load, shear load and bending axes. The hybrid method procedure is as follows:

(a) Failure tests are performed with respect to loading conditions.
(b) Finite element simulations are performed for each experiment.
(c) Based on the failure loads obtained in failure test, the instant of onset of spot weld failure is determined. Failure loads are extracted comparing experiments with simulations.
(d) Post-processing of those simulations gives failure load components acting on spot welds such as normal, shear and bending loads.
(e) These failure load components are plotted on the plane consisting of normal, shear and bending axes.

Figure 9.4 shows the hybrid method to obtain the failure load components from the failure tests.

9.5 Construction of Newly Proposed Failure Model

Based on the hybrid method, newly proposed failure model is constructed by Eq. (9.5) as shown in Fig. 9.5. The coefficients that constitute newly proposed failure model are obtained using the least square method to minimize the discrepancy between the experimental data and interpolated data.

Generally, the surface of the conventional spot weld failure models is convex shape. However, the shape of failure surface is changed to material strength. As the material is stronger, the moment effect to spot weld failure is significant. In this paper, materials with various levels of strength were tested to develop the practical spot weld failure model as shown in Fig. 9.6. The shape of failure surfaces change from convex to concave as the material strength increases. Figure 9.7 shows failure surfaces of the proposed model and that of the conventional model. Conventional models tend to evaluate the spot weld fracture to be excessively safe. Therefore, newly proposed failure model have to be utilized in the crash simulation to predict the spot weld failure.

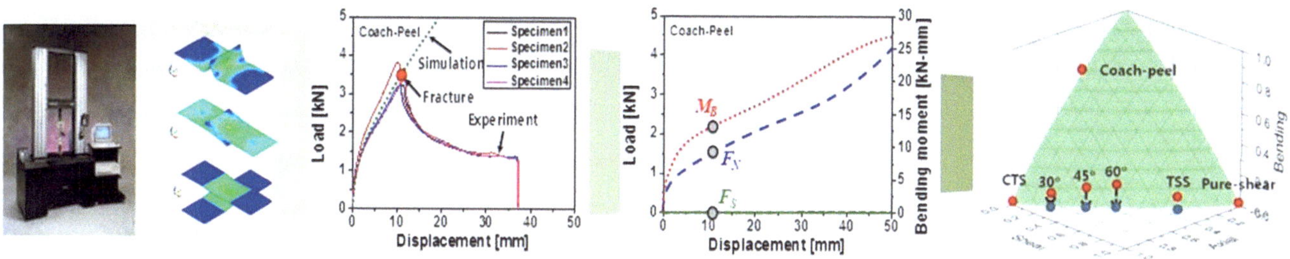

Fig. 9.4 Hybrid method to obtain the failure load with respect to test conditions. *1* test, *2* simulation, *3* extract fracture load, *4* calculate load components, *5* build fracture surface

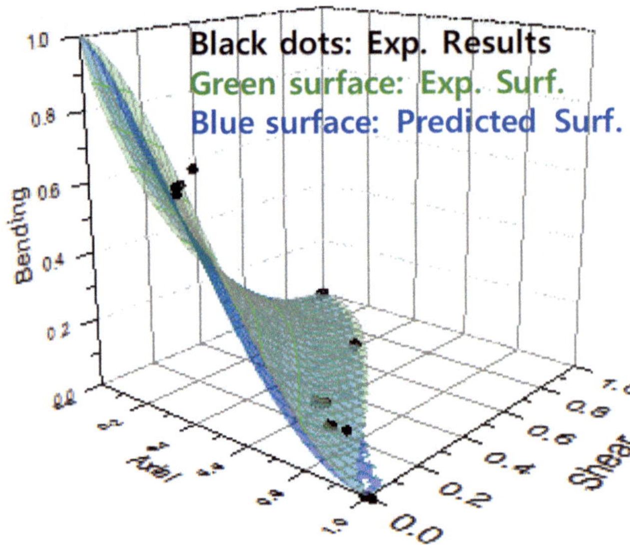

Fig. 9.5 Spot weld failure surface constructed by newly proposed failure model

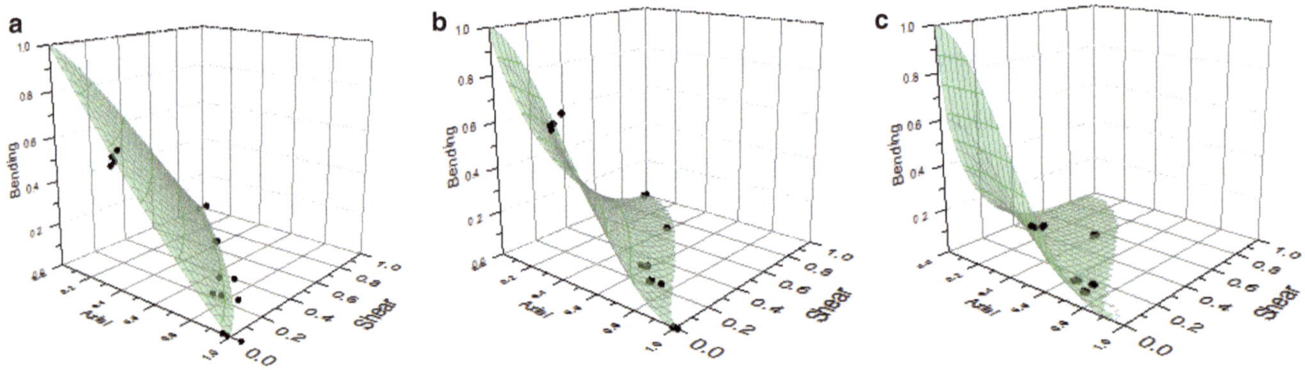

Fig. 9.6 Failure surface of spot welds with respect to the material grades: (a) DQ; (b) 440R; (c) 780DP

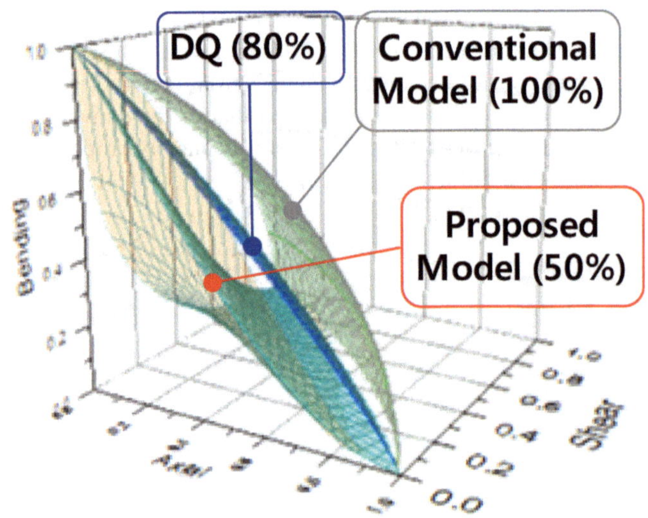

Fig. 9.7 Failure surfaces of the proposed model and the conventional model

The proposed failure model for spot weld in this paper can predict the spot weld failure accurately. However, a number of failure tests and analysis have to be conducted to construct the failure model. In order to construct the failure model simply, prediction equations were developed for the six coefficients of the proposed spot weld failure model. These equations are expressed as

$$F_n, F_s, M_b = f(t, \sigma_{TS}, \phi) \quad (9.6)$$

$$\alpha, \beta, \gamma = f(t, \sigma_{TS}, \phi) \quad (9.7)$$

Here, t, σ_{TS} and ϕ are thickness, tensile strength and nugget ratio, respectively. six coefficients are determined by the function of thickness, tensile strength and nugget ratio. Blue failure surface shown in Fig. 9.5 is constructed by the prediction equations. Predicted failure surface describes the failure surface constructed by failure tests and hybrid method.

9.6 Verification of the Proposed Failure Model

Finite element analyses of the cross tension, loading angle of 30°, loading angle of 45°, coach-peel and lap-shear tests were conducted considering the failure of a spot weld. The spot welded part is modeled with a cluster of 16 hexahedron elements which connects a pair of shell elements by constraining all of the degrees of freedom. The mesh size of the shell elements was set to 1 mm in order to describe the deformation of a steel sheet around the spot weld. The failure model of 440R 1.2 t steel sheet is expressed in Eq. (9.8).

$$\left(\frac{f_n}{9.27}\right)^2 + \left(\frac{f_s}{19.79}\right)^{0.723} + \left(\frac{m_b}{17.30}\right)^{0.5} = 1 \quad (9.8)$$

In order to find the failure point of spot welds in FE simulation, spot weld failure assessment system (SWFAS) was developed based on the newly proposed failure model and LS-DYNA result file such as SWFORC to confirm the failure spot welds from the crash simulation results. SWFAS program informs the failure location of spot welds and the degree of failure risk to SWFAS program users. SWFAS program assesses spot weld failure by LS-DYNA result file, SWFORC, without additional simulation.

Figure 9.8 compares load-displacement curves obtained from the analyses to those from experiments. The failure loads obtained from the experiments are summarized in Table 9.1. The comparison represents that the failure loads of the spot weld obtained from the analyses are close in coincidence with those obtained from the experiments. When the failure tests were performed using jig system such as the cross tension test and the loading angle imposed tests, differences are observed between the experimental results and the simulation results. This phenomenon is observed because of slip between specimen and testing jig. Proposed failure model can predict the failure location and the degree of failure risk. Therefore, proposed failure model and SWFAS program have to be used in order to predict the failure of spot weld in crash simulations.

9.7 Conclusion

A number of automotive steel sheets are evaluated and analyzed by various spot-weld tests. In order to develop a failure model of spot welds, a new methodology is suggested to obtain the load components of spot weld based on the combination of experiments and simulation which is called a hybrid method. The failure surface is determined based on bending moment, normal force and shear force. The shape of failure criterion to material strength is changed. As the material is stronger, the moment effect to spot weld failure is significant. The failure surface becomes concave and sharp as the material strength increases. Finite element analyses of specimens are performed considering the failure of a spot weld. The failure loads obtained from the analyses are close in coincidence with those obtained from the experiments. It can be concluded that the failure model is appropriate to describe the failure behavior of spot welds in vehicle simulations.

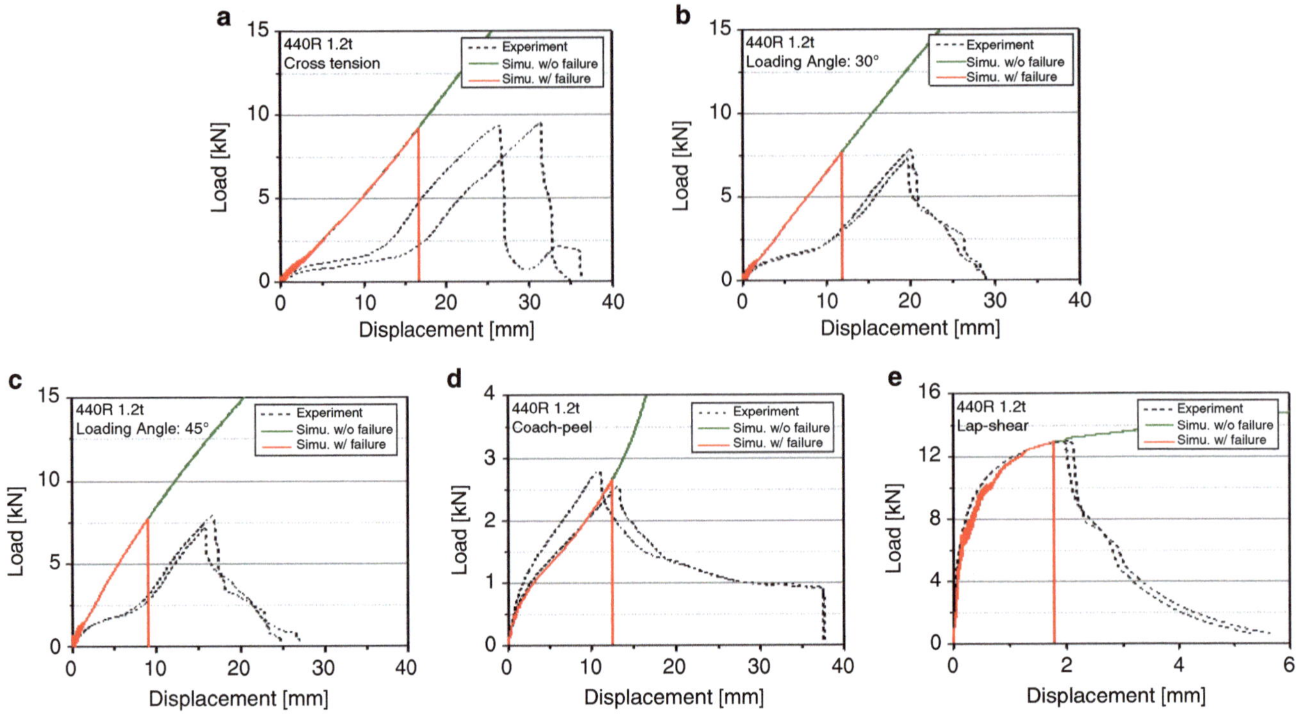

Fig. 9.8 Comparison of the load-displacement curves from the FE analyses and the experiments: (**a**) cross tension; (**b**) loading angle of 30°; (**c**) loading angle of 45°; (**d**) coach-peel; (**e**) lap-shear

Table 9.1 Comparison of the failure load obtained from the FE analyses and the experiments

	Failure load [kN]			
Test condition	Exp. 1	Exp. 2	Simul.	Error (%)
Cross tension test	9.53	9.42	9.23 (1.00)	~3.1
Loading angle of 30°	7.84	7.31	7.71 (1.00)	~5.5
Loading angle of 45°	7.20	7.90	7.74 (1.00)	~7.5
Coach-peel test	2.53	2.79	2.65 (1.00)	~5.0
Lap-shear test	12.94	12.97	13.00 (1.00)	~0.5

References

1. Wung, P.: A force-based failure criterion for spot weld design. Exp. Mech. **41**(1), 107–113 (2001)
2. Wung, P., Walsh, T., Ourchane, A., Stewart, W., Jie, M.: Failure of spot welds under in-plane static loading. Exp. Mech. **41**(1), 100–106 (2001)
3. van Schaik, M., Martin, D.C., Denner, S.: ULSAB advanced vehicle concepts—the latest steel demonstration for automotive. SAE 2000-01-1545 (2000)
4. Sato, K., Yoshitake, A., Hosoya, Y., Mikami, H.: FEM simulation to estimate crashworthiness of automotive. SAE 982356 (1998)
5. Inoue, T., Nakanishi, E.: A crash simulation analysis which consider the SPOTWELD rupture. JSAE, no. 88-00, pp. 5–7 (2000)
6. Zhang, H., Senkara, J.: Resistance Welding–Fundamentals and Application. CRC Press, Florida (2006)
7. Lee, Y.L., Wehner, T.J., Lu, M.W., Morrissett, T.W., Pakalnins, E.: Ultimate strength of resistance spot welds subjected to combined tension and shear. J. Test. Eval. **26**(3), 213–219 (1998)
8. Barkey, M.E., Kang, H.: Testing of spot welded coupons in combined tension and shear. Exp. Tech. **23**(5), 20–22 (1999)
9. Madasamy, C., Tyan, T., Faruque, O., Wung, P.: Methodology for testing of spot-welded steel connections under static and impact loadings. SAE 2003-01-0608 (2003)
10. Langrand, B., Combescure, A.: Non-linear and failure behavior of spotwelds: a "global" finite element and experiments in pure and mixed modes I/II. Int. J. Solids Struct. **41**, 6631–6646 (2004)

11. Lin, S.-H., Pan, J., Wu, S.-R., Tyan, T., Wung, P.: Failure loads of spot welds under combined opening and shear static loading conditions. Int. J. Solids Struct. **39**, 19–39 (2002)
12. Lin, S.-H., Pan, J., Tyan, T., Prasad, P.: A general failure criterion for spot welds under combined loading conditions. Int. J. Solids Struct. **40** (21), 5539–5564 (2003)
13. Seeger, F., Feucht, M., Frank, Th., Keding, B., Haufe, A.: An investigation on spot weld modeling for crash simulation with LS-DYNA. In: Proceeding of the 4th German LS-DYNA Forum, Bamberg, Germany, B-I-1-B-I-12 (2005)
14. Seeger, F., Michel, G., Blanquet, M.: Investigation of spot weld behaviour using detailed modeling technique. In: Proceeding of the 7th LS-DYNA Anwenderforum, Bamberg, Germany, B-I-29-B-I-38 (2008)
15. Yang, Y.-P., Orth, F., Peterson, W., Gould, J.: Accurate spot weld testing for automotive applications. Adv. Mater. Process. **172**(11), 19–22 (2014)
16. Sandahl, A., Jergeus, J., Centeno, O., Macri, D., Ericsson, A., Wu, W., Claesson, E., Eggertsen, P.A., Retzlaff, M., Khoo, M.: Crash CAE in the all new Volvo XC90 and SPA platform. In: Proceeding of the 10th European LS-DYNA Conference, Würzburg, Germany (2015)
17. Chao, Y.: Ultimate strength and failure mechanism of resistant spot weld subjected to tensile, shear, or combined tensile/shear loads. ASME J. Eng. Mater. Technol. **125**, 125–132 (2003)
18. Ha, J., Huh, H.: Failure characterization of laser welds under combined loading conditions. Int. J. Mech. Sci. **69**, 40–68 (2013)
19. Dunand, M., Mohr, D.: Hybrid experimental–numerical analysis of basic ductile fracture experiments for sheet metals. Int. J. Solids Struct. **47**, 1130–1143 (2010)
20. Huh, H., Lim, J.H., Park, S.H.: High speed tensile test of steel sheets for the stress-strain curve at the intermediate strain rate. Int. J. Autom. Technol. **10**(2), 195–204 (2009)
21. Ha, J.W., Song, J.H., Huh, H., Lin, J.H., Park, S.H.: Dynamic material properties of the heat-affected zone (HAZ) in resistance spot welding. Int. J. Modern Phys. B **22**(31/32), 5800–5806 (2008)

Chapter 10
Constitutive Model Calibration via Autonomous Multiaxial Experimentation

P.L. Phillips, R.A. Brockman, D.J. Buchanan, and R. John

Abstract Modern plasticity models contain numerous parameters that can be difficult and time consuming to fit using current methods. Additional experiments are seldom conducted to validate the model for experimental conditions outside those used in the fitting procedure. To increase the accuracy and validity of these advanced constitutive models, software and testing methodology have been developed to seamlessly integrate experimentation, parameter identification, and model validation in real-time over a range of multiaxial stress conditions, using an axial/torsional test machine. Experimental data is reduced and finite element simulations are conducted in parallel with the test based on experimental strain conditions. Optimization methods reconcile the experiment and simulation through changes to the plasticity model parameters. Excursions into less-traveled portions of the multiaxial stress space can be predicted, and then executed experimentally, to identify deficiencies in the model. Most notably, the software can autonomously redirect the experiment to increase the robustness of the plasticity model where further deficiencies are identified, thus providing closed loop control of the experiment. This novel process yields a calibrated plasticity model upon test completion that has been fit and more importantly validated, and can be used directly in finite element simulations of more complex geometries.

Keywords Axial/torsional experimentation • Plasticity • Constitutive model fitting • Autonomous testing • FEMU

10.1 Introduction

Due to the high cost of designing and manufacturing complex subcomponents, it is becoming common for finite element analysis to replace experimental tests to avoid the costs associated with elaborate test fixtures and equipment or expensive physical specimens. In addition, certain stress states commonly of interest are difficult, if not impossible, to physically reproduce in experiments. Accurate material models that are calibrated over a wide range of stress and temperature conditions are critical to ensuring the accuracy and reliability of finite element results for complicated geometries subjected to realistic loading environments. Identifying parameters for modern constitutive models, that continue to grow in complexity, is a difficult and time consuming process that has historically be a separate process from the experimental testing. As such, additional experiments are seldom conducted to validate the model for experimental conditions outside those used in the fitting procedure.

Current methods for parameter identification are either separate endeavors as is the case with Finite Element Model Updating (FEMU) [1, 2] or are still limited in the scope of nonlinear models or stress states as is the case with the growing area of the Virtual Fields Method (VFM) [3]. Multiaxial experimentation [4–6], specifically axial/torsional testing [7], is a mature technology and modern equipment is capable of testing at various rates and elevated temperatures; however, material testing is still dominated by tension testing. Finite element methods continue to advance and the constitutive models (including modern plasticity models) used within continue to evolve and grow in complexity as researchers seek to describe

P.L. Phillips (✉) • R.A. Brockman • D.J. Buchanan
University of Dayton Research Institute, 300 College Park, Dayton, OH 45409-0110, USA

Air Force Research Laboratory (AFRL/RXCM), Wright-Patterson AFB, OH 45433-7817, USA
e-mail: Peter.Phillips@udri.udayton.edu

R. John
Air Force Research Laboratory (AFRL/RXCM), Wright-Patterson AFB, OH 45433-7817, USA

Y. Zhu, A.T. Zehnder (eds.), *Experimental and Applied Mechanics, Volume 4*,
Conference Proceedings of the Society for Experimental Mechanics Series, DOI 10.1007/978-3-319-42028-8_10

more complicated behavior, e.g. low-cycle fatigue, non-proportional loading, etc. For an extensive review of modern plasticity models, see Chaboche [8]. Computer hardware has reached the point where multiple CPU systems are relatively inexpensive and parallel processing architectures are readily available in OpenMP and MPI. The maturity of the aforementioned fields enables the advancement of the state-of-the-art in experimental testing through the combination of multiaxial experiments with real-time constitutive model calibration. Software and testing methodology have been developed to seamlessly integrate experimentation, parameter identification, and model validation in real-time over a range of multiaxial stress conditions, using an axial/torsional test machine. The methodology is material agnostic; however, initial work has been focused on initially isotropic materials using phenomenological plasticity models. Upon test completion, this new, novel methodology yields a calibrated plasticity model that has been fit and more importantly validated, and can be used immediately in finite element simulations of more complex geometries. The new method also reduces the time required to perform the complex fitting and decreases the required number and complexity of the test specimens. It also significantly reduces the time to develop a validated model since the experiments are done as needed to fit the model and in real-time as opposed to post-test.

10.2 Methodology

Many different pieces of software are required to perform constitutive model calibration through autonomous multiaxial testing. Figure 10.1 provides an overview of the entire methodology and illustrates how the different pieces of software communicate with each other. Experimental data is reduced and finite element simulations are conducted in parallel with the test based on experimental strain conditions. Optimization methods reconcile the experiment and simulation through changes to the plasticity parameters. Excursions into less-traveled portions of the multiaxial stress space are predicted by the testing platform, and then executed experimentally, to identify deficiencies in the model. Most notably, the software can autonomously redirect the experiment to increase the robustness of the plasticity model where further deficiencies are identified, thus providing closed loop control of the experiment.

The model fitting relies on Finite Element Model Updating (FEMU) which requires both a finite element code and an optimizer. Developing autonomous experimental control of the axial/torsional testing hardware requires software for test control, acquisition of data, and communication with the external software responsible for the overall integration. Section 10.2.1 describes the axial/torsional testing machine and test specimen details. Section 10.2.2 describes the

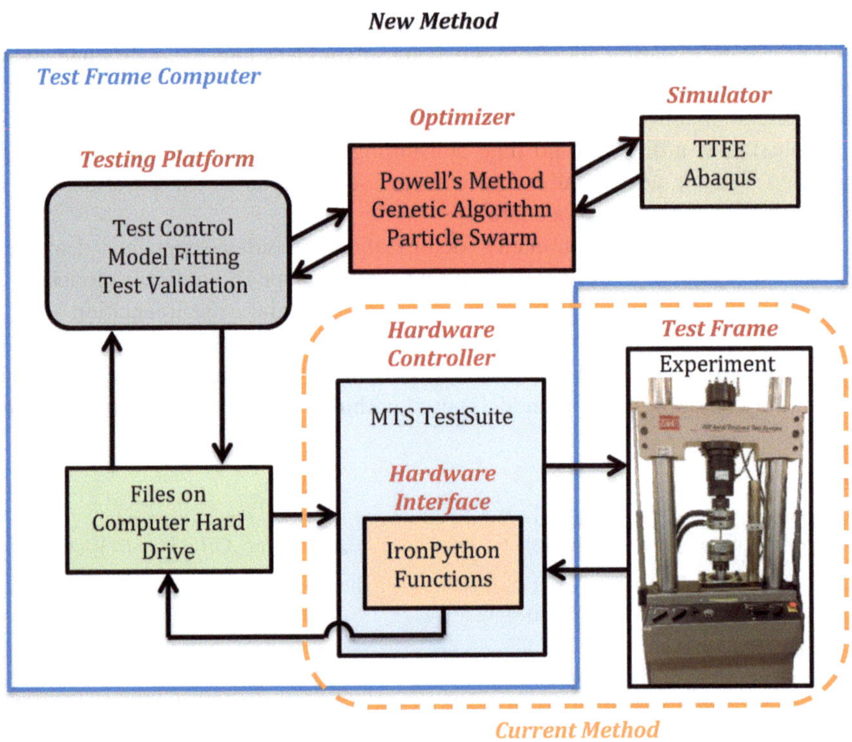

Fig. 10.1 Overview of the new software and test methodology that allows for autonomous multiaxial experimentation and constitutive model fitting

specialized finite element code written for its compact size, fast execution, and integration with the optimization program. Section 10.2.3 describes the optimization program and the different methods available. Lastly, Sect. 10.2.4 outlines the new novel software responsible for the autonomous constitutive model calibration via multiaxial experimentation.

10.2.1 Experimental Setup

The experiments were conducted on a MTS 809 Axial/Torsional test frame with force and torque capabilities of 100 kN and 1100 N m, respectively. The test frame was controlled using a custom program written in MTS TestSuite. Axial and shear strain measurements were obtained using a MTS high-temperature axial/torsional extensometer with a gage length of 25.0 mm. The specimen used for the initial capability demonstrations was made from 4340 steel, with dimensions as shown in Fig. 10.2.

10.2.2 Simulator (Finite Element Program)

A finite element program has been written that is customized for isotropic materials subjected to axial/torsional loading. This allows for accurate simulation of multiaxial loading for both solid and thin-walled cylindrical specimens made of materials with continuum level isotropic behavior. The code uses axisymmetric, four node elements formulated in cylindrical coordinates with displacement degrees of freedom in the radial, circumferential, and axial directions. Figure 10.3 shows the simplification from the solid three-dimensional mesh to the axisymmetric mesh used in the code. All three degrees of freedom are assumed to be independent of the angular position; therefore, the four node elements are modeled in the theta equals zero plane. The program handles material nonlinearity and implements the material routines through Abaqus/Standard [9] user material subroutines (UMAT) that are written in Fortran. The same material subroutines used in the custom finite element code are directly compatible with the commercial code, Abaqus. Material models for elastic and linear isotropic hardening are also included.

A UMAT has been written for a unified plasticity model with combined nonlinear isotropic-kinematic hardening. The development of the rate equations and the consistent material tangent matrix follows Kirchner [10] with the exception that the yield function has been simplified to include the von Mises stress. This Chaboche type plasticity model has been used successfully by others [8] for the simulation of cyclic experiments. The rate equations describing the model are

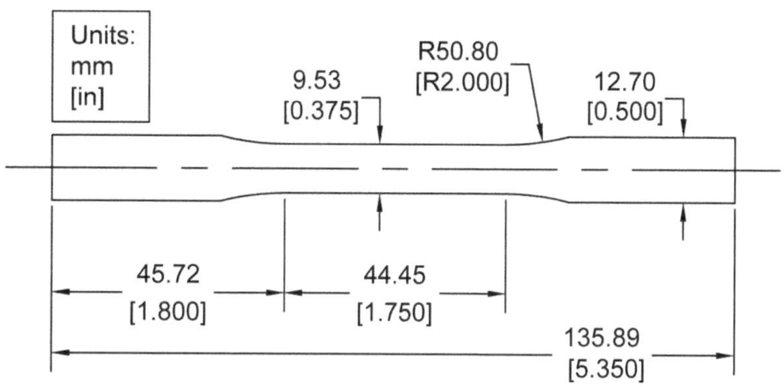

Fig. 10.2 Solid cylindrical specimen made from 4340 steel

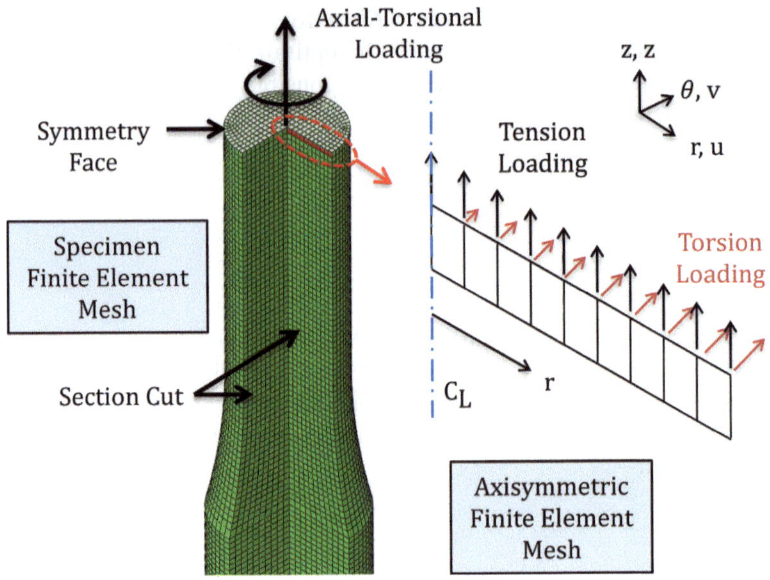

Fig. 10.3 Simplification of a solid cylindrical specimen from three-dimensional solid elements to the axisymmetric elements in polar coordinates used in the custom finite element software

$$
\begin{aligned}
f &= \sqrt{\frac{3}{2}\left(S_{ij}-X_{ij}\right)\left(S_{ij}-X_{ij}\right)} - \rho_0 - \sum_{w=1}^{N}\rho^{(w)} \\
\dot{\sigma}_{ij} &= D_{ijkl}\left(\dot{\varepsilon}_{kl} - \left\langle\frac{f}{D}\right\rangle^{n}\frac{\partial f}{\partial\sigma_{kl}}\right) \\
\dot{\rho}^{(w)} &= b^{(w)}\left(Q^{(w)}-\rho^{(w)}\right)\left\langle\frac{f}{D}\right\rangle^{n} \\
\dot{\chi}_{ij}^{(w)} &= \left(\frac{2}{3}A^{(w)}\frac{\partial f}{\partial\sigma_{ij}} - B^{(w)}\chi_{ij}^{(w)}\right)\left\langle\frac{f}{D}\right\rangle^{n} \\
X_{ij} &= \sum_{w}\chi_{ij}^{(w)} - {}^{1}\!/_{3}\,\delta_{ij}\chi_{kk}^{(w)}
\end{aligned}
\qquad (10.1)
$$

where f is the yield function, ρ_0 is the initial yield surface size, $\rho^{(w)}$ are the individual isotropic hardening terms, D is the drag stress, n is the rate sensitivity exponent, and $\chi_{ij}^{(w)}$ are the individual backstress terms. The yield function uses the deviatoric stresses, S_{ij}, and backstresses, X_{ij}. The backstresses have both a linear hardening term, A, and a recovery term, B. The isotropic hardening terms have a saturation stress, Q, and a rate term, b. In addition a series of terms can be used for both the isotropic and kinematic terms to achieve better calibration over multiple strain ranges.

10.2.3 Optimizer

A critical part of the methodology developed is the optimization software as this directly influences to success of the constitutive model fitting. When using FEMU, each objective function evaluation requires a separate call to the finite element program with a different set of material parameters. The optimization seeks to minimize the objective function given in Eq. (10.2), where N is the number of load steps (experimental data points) used in the FE solution, and k denotes the individual load step. The first and second terms of the objective function are related to the force, F, and torque, T, respectively. The subscript 'exp' denotes experimental results while 'FE' denotes finite element results. Each term has a weight factor that can reconcile the difference in magnitude between the force and torque. For this work, w_F is A^{-2} where A is the initial cross-section area, and w_T is r^2J^{-2} where r is the outer radius of the specimen and J is the polar moment of inertia. With the chosen weight factors, the objective function has units of stress squared. The experimental values come directly from the load cell, while the FE values are obtained through integration of element stresses.

$$obj = \sum_{k=1}^{N} w_F \left(F_{\exp}^{(k)} - F_{FE}^{(k)} \right)^2 + w_T \left(T_{\exp}^{(k)} - T_{FE}^{(k)} \right)^2 \tag{10.2}$$

The optimization software currently has three methods available: Powell's method, a genetic algorithm (GA), and particle swarm optimization (PSO). The three methods have shown to yield fairly similar results when similar inputs and tolerances are used; therefore, the GA and PSO are used as the objective functions can be evaluated in parallel.

The GA is a simple version with concepts taken from Goldberg [11]. The GA uses bound constraints, tournament selection with three parents, uniform crossover, and mutation. Ten parents are used, the probability of crossover and mutation is 0.5 and 0.02, and ten rounds are carried out for a total of 100 objective function evaluations. The PSO has inertia, cognitive, and social factors similar to Kennedy [12] with an additional pheromone velocity factor similar to Foo et al. [13]. The parameter values are bounded through a simple rebound formula in which a particle that exceeds the bounds is reflected at the boundary and the reflected portion loses half its magnitude. Ten particles are used for ten rounds with velocity factors ramped linearly from starting values to end values over the specified number of rounds. Initial and final value pairs for the inertia, cognitive, social factors, and pheromone factors are (0.5, 0.5), (1.0, 2.0), (2.0, 1.0), and (1.0, 1.0) respectively. The initial populations for both methods have been formulated using a Latin-hypercube sampling method similar to Singh et al. [14]. Initial velocities for the PSO are equated to the difference between a second Latin-hypercube population on the initial population. Lastly, velocity clamping is used for the PSO to limit the maximum Euclidean norm of the velocity to 0.33.

10.2.4 Testing Platform

The testing platform software includes the optimizer, simulator, methods to communicate with the MTS test frame through the hardware controller, and additional intelligent algorithms to perform the autonomous multiaxial experimentation and constitutive model fitting. Experimental control is determined on-the-fly rather than requiring the entire process to be predefined because the testing platform software runs simultaneously with the experiment, collects the experimental data, processes the data, and dictates the next loading step for the experiment. The loading steps can be a single ramp from one load level to another, a series of cycles between prescribed load levels, or dwelling at current loads. Control algorithms are included in the testing platform to perform: initial modulus checks for signal verification, zero-load offsets for strain channels based on least-squares regression, yield surface probing for multiple stress trajectories using single specimens (similar to Lissenden et al. [15]), and most importantly, autonomous constitutive model fitting and multiaxial experimentation.

The software can accommodate control modes for load, stroke, and strain implemented through stroke (pseudo-strain). The load and stroke control utilize the built-in capabilities of the test frame controller. However; the pseudo-strain control requires an understanding of the relationship between increments in stroke and corresponding strain increment. This correlation is determined during the initial modulus check. In addition, calculations are done to map the desired strain level to a required stroke, and limit detection is used to stop the load excursion once the requested strain level is reached. As plasticity occurs, the stroke limit predictions based on correlation from elastic deformation will be excessive; however, the strain limit detection stops the stroke at the desired strain. The pseudo-strain method is implemented to bypass the possible instabilities of strain control, and to also allow for the possibility of redundant strain control mechanism (if the extensometer fails or jumps, strain gage data could be used as the controlling signal).

Prior to use in the optimization software, the experimental data is filtered using a lowpass filter. Both eighth-order Butterworth and 100-pole finite impulse response (FIR) filters are available within the software with various cutoff frequencies. The default used is the 100-pole FIR filter with a 10 Hz cutoff. Each loading segment of the filtered data is then further down sampled based on the elastic modulus and user-defined increments in strain over the nonlinear region. This reduces the number of load increments required for the finite element analysis.

The software needs to be initialized with information regarding the desired material model and definitions of the testing boundaries with respect to the axial and shear stress or strain space. Each constitutive model has its own setup routine where the user can prescribe which parameters of the model they wish to optimize. In addition, they can set bounds on the parameter values or accept the default values present in the software. The testing bounds are described in terms of the maximum and minimum levels, or maximum levels and R-ratios where R is the maximum over the minimum load/strain. A sequence of rounds can be prescribed in which the maximum levels increase over each round. Figure 10.4 shows an example for a pseudo-strain-controlled test in which the maximum strain limits were 1.0 % strain. This example also

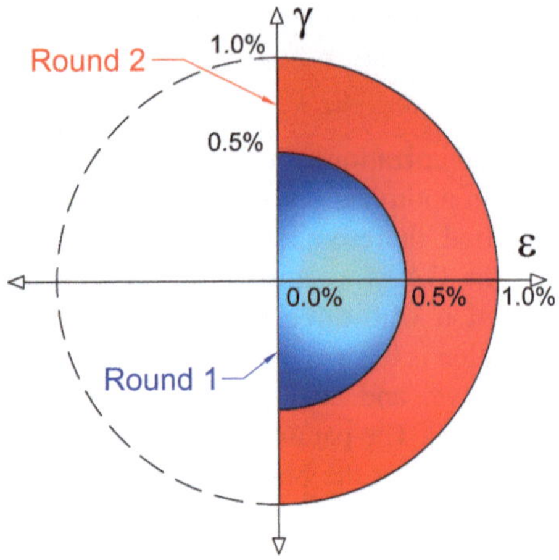

Fig. 10.4 Test conditions for autonomous model fitting with: strain control, ±90° with no compressive strain, and two rounds with a maximum strain of 0.5 and 1.0 %

corresponds to the results presented in the next section. Two rounds are chosen, with the test bounds between ±90° with no compressive strain allowed. The behavior within each round can be predefined by the user or randomly assigned by the software, and each round can consist of multiple stress trajectories within the defined test bounds.

The autonomous fitting procedure directs the experimental test based on the setting for the current round. Force, torque, and strain data from the experiment are read by the software, filtered, and down sampled for use in the optimization program. Some initial portion of the experimental data can be used in the model fitting process, and the remainder can be used for model validation. Plasticity models are history dependent so the entire loading process must be simulated whenever the model parameters are revised. As an example, if the test has a total of three rounds, the first two rounds can be used for the model fitting and the third round can be used for validation to determine the accuracy of the optimized material parameters. The criterion for which data is used in the prediction and which is used for validation is still evolving and will be the focus of future work. In addition, the criteria for when the autonomous fitting procedure has deemed itself converged is continuing to evolve.

10.3 Results

Testing was carried out on a solid cylindrical specimen (Fig. 10.2) made from 4340 steel. Force and torque measurements were obtained from the 100 kN, 1100 N m load and torque cells. Both axial and shear strain data were collected using the axial/torsional extensometer. Prior to starting the fitting procedure, the axial and shear moduli were determined as well as parameters required for using the pseudo-strain algorithm. Rather than perform the yield surface probing, a yield stress of 100 MPa was assumed (based on yield surface probing of a separate specimen). The autonomous model fitting procedure was performed under pseudo-strain control for two rounds with maximum strain levels of 0.5 and 1.0 % strain. Figure 10.4 depicts the test conditions. At each strain level, the specimen underwent: axial only loading (tension only), fully reversed torsional loading, and combined axial and torsional load (again tension only). Figure 10.5 shows experimental results for engineering stress vs. engineering strain for both axial and shear stresses. Both rounds of strain levels were used for the prediction of the material properties. Development of more advanced algorithms for the model fitting and validation data set definitions will continue to evolve as more experience is gained in working with the new software.

The plasticity model with nonlinear isotropic and kinematic hardening given in Eq. (10.1) was used during the modeling fitting procedure. A single isotropic hardening term was used (with two parameters b and Q), and two nonlinear kinematic hardening terms were used. The first kinematic term used both parameters A1, and B1, while the second kinematic term only used the linear parameter A2, with the corresponding recovery parameter B2 set to zero. The optimization was therefore performed for a total of five material parameters (b, Q, A1, B1, and A2). Both the GA and PSO were used for comparison of the methods with settings as described in the *Optimizer* section.

Table 10.1 provides results for the optimized material parameters, and also provides computation times for each algorithm. Both algorithms were run using two computer processors (thus two FE jobs are run simultaneously). Preliminary scalability testing of the parallelization has shown a scalability factor of 0.92 out of 1.0. Both optimization methods converged on similar material parameters, and the objective function values are in reasonable agreement. Figure 10.6 compares the stresses from the experiment with the stresses from the best set of parameters resulting from the GA and PSO.

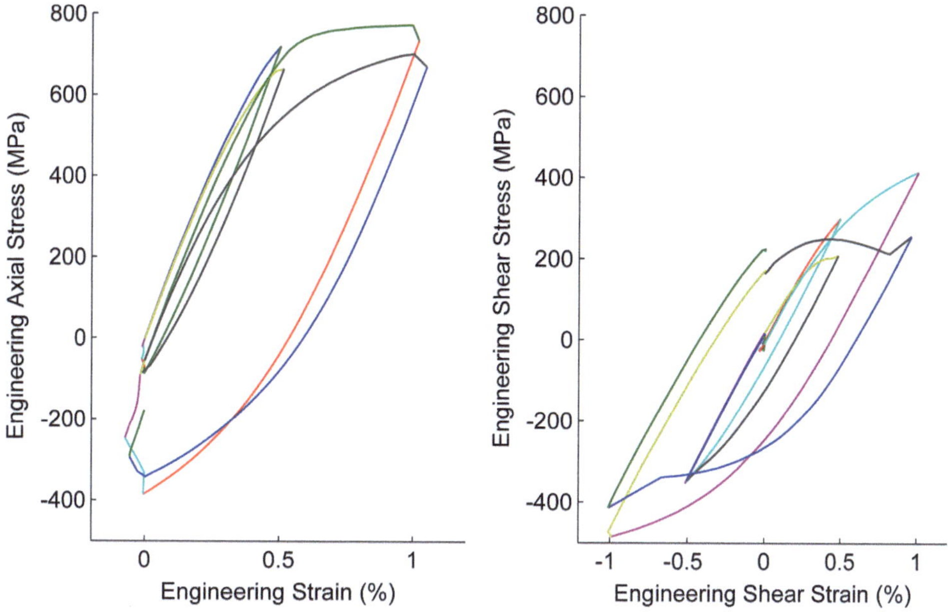

Fig. 10.5 Strain controlled experimental results for solid cylindrical specimen made of 4340

Table 10.1 Material property fitting optimization results for both GA and PSO

Method	Computer cores	Number of analyses	Run time (min)	Run time (s)	Obj fun.
GA	2	100	23.6	1414	4,373,610
PSO	2	100	16.2	969	3,607,640
	Par	**GA**	**PSO**	**min**	**max**
	Q	250	224	10	250
	b	5520	4550	500	9000
	A1	290,370	108,160	50	9,000,000
	A1/B1	297	341	10	1000
	B1	976	317	5	9000
	A2	3,226,500	5,846,300	1000	10,000,000
	B2	–	–		

The optimized set of parameters follow the same trends as the experiment, but there is room for improvement. Changes to the control settings of the optimization methods will likely reduce the difference in peak magnitudes between the experimental and FE results. Work is ongoing to improve the optimization control settings.

10.4 Conclusions

To increase the accuracy and validity of modern plasticity models, software and testing methodology have been developed to seamlessly integrate experimentation, parameter identification, and model validation in real-time over a range of multiaxial stress conditions, using an axial/torsional test machine. This novel process can yield a calibrated plasticity model upon test completion that has been fit and more importantly validated, and can be used directly in finite element simulations of more complex geometries. This work is a significant step in advancing how materials are tested in the laboratory and characterized using constitutive models. The new methods can produce calibrated models in real-time for multiaxial conditions (rather than the predominate trend of uniaxial testing) and will likely require less time and fewer specimens. The methodology is material independent and the finite element software and material models could be extended to anisotropic materials if desired. Future work will focus on improving the fitting and validation algorithms, and applying the new testing methodology to materials of interest in the aerospace community, namely nickel-based superalloys and titanium.

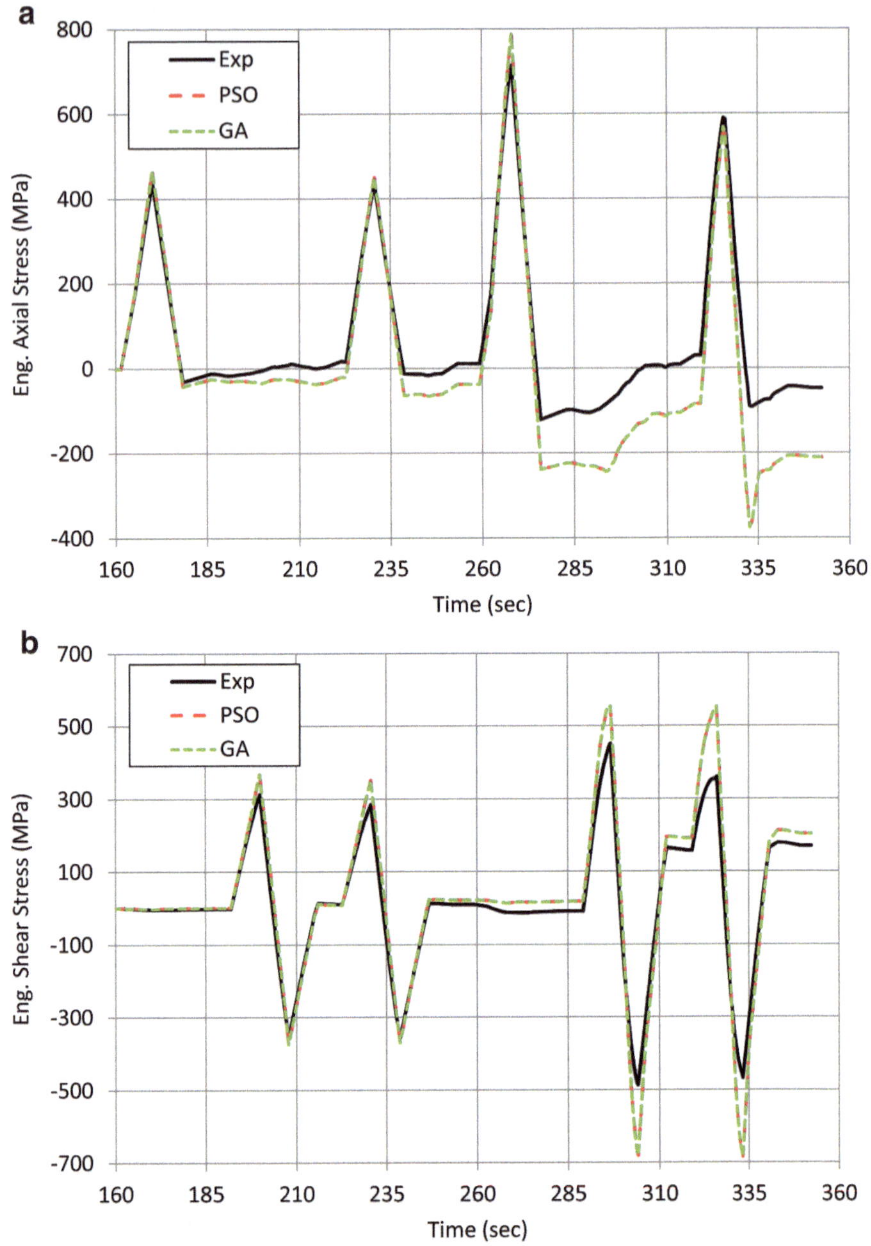

Fig. 10.6 Comparison of the (**a**) axial stress and (**b**) shear stresses from the experiment (*solid line*) and the nearly identical results from the GA and PSO optimization (both *dashed*)

Acknowledgements This work has been sponsored through the AFRL/DAGSI Ohio Student-Faculty Research Fellowship program, Topic RX14-12. Additional funding assistance was also provided through the Air Force Research Laboratory, AFRL/RXCM, Wright-Patterson Air Force Base, OH, under DoD contract FA8650-14-D-5205.

The authors would like to thank the Air Force Research Laboratory, AFRL/RXCM for the use of their axial/torsional testing equipment and Mr. Philip E. Blosser for his assistance in operating the equipment.

References

1. Fedele, R., Filippini, M., Maier, G.: Constitutive model calibration for railway wheel steel through tension–torsion tests. Comput. Struct. **83**(12–13), 1005–1020 (2005)
2. Avril, S., Pierron, F., Sutton, M.A., Yan, J.: Identification of elasto-visco-plastic parameters and characterization of Lüders behavior using digital image correlation and the virtual fields method. Mech. Mater. **40**(9), 729–742 (2008)
3. Pierron, F., Grédiac, M.: The Virtual Fields Method. Springer, New York (2012)
4. Guest, J.J.: On the strength of ductile materials under combined stress. Proc. Phys. Soc. Lond. **17**(1), 202 (1899)

5. Osakada, K.: History of plasticity and metal forming analysis. J. Mater. Process. Technol. **210**(11), 1436–1454 (2010)
6. Michno Jr., M.J., Findley, W.N.: An historical perspective of yield surface investigations for metals. Int. J. Non Linear Mech. **11**(1), 59–82 (1976)
7. Cailletaud, G., Kaczmarek, H., Policella, H.: Some elements on multiaxial behaviour of 316 L stainless steel at room temperature. Mech. Mater. **3**(4), 333–347 (1984)
8. Chaboche, J.L.: A review of some plasticity and viscoplasticity constitutive theories. Int. J. Plast. **24**(10), 1642–1693 (2008)
9. Dassault Systemes: ABAQUS User's Manual, Version 6.13. Dassault Systemes Inc. (2013)
10. Kirchner, E.: Modeling single crystals: time integration, tangent operators, sensitivity analysis and shape optimization. Int. J. Plast. **17**(7), 907–942 (2001)
11. Goldberg, D.E.: Genetic Algorithms in Search, Optimization, and Machine Learning, 30. Print. Addison-Wesley, Boston (2012)
12. Eberhart, R.C., Kennedy, J.: A new optimizer using particle swarm theory, In: Proceedings of the Sixth International Symposium on Micro Machine and Human Science, vol. 1, pp. 39–43 (1995)
13. Foo, J.L., Kalivarapu, V., Winer, E.: Implementation of digital pheromones for use in particle swarm optimization (2006)
14. Singh, G., Spradlin, T.J., Grandhi, R.V.: Fatigue life optimization using a laser peening process, In Proceedings of the 50th AIAA/ASME/ASCE/AHS/ASC Structures, Structural Dynamics, and Materials Conference, Palm Springs, California (2009)
15. Lissenden, C.J., Lerch, B.A., Ellis, J.R., Robinson, D.N.: Experimental determination of yield and flow surfaces under axial-torsional loading. ASTM Spec. Tech. Publ. **1280**, 92–112 (1997)

Chapter 11
Measurement of Structural Stresses by Hole-Drilling and DIC

Joshua Harrington and Gary S. Schajer

Abstract Measurement of stresses in structures such as bridges, buildings, pipelines and railways is difficult because the loads cannot easily be manipulated to allow direct measurements. This paper focuses on the development of a method that combines the hole-drilling technique with Digital Image Correlation (DIC) to evaluate these difficult-to-measure structural stresses. The hole-drilling technique works by relating local displacements caused by the removal of a small amount of stressed material to the original stresses within the drilled hole. Adaptation of this method to measure structural stresses requires scaling up the hole size and modifying the calculation approach to measure deeper into a material. DIC provides a robust means to measure full-field displacements that can easily be scaled to different hole sizes and corrected for measurement artifacts. There are two primary areas of investigation: the adaptation of the DIC/hole-drilling method to measure structural stresses and the development of a correction method to remove coexisting stresses such as residual and machining stresses from the measurement. Experimental measurements are made to demonstrate the measurement method on different structure types including the example practical problem of measuring thermally induced stresses in railroad tracks.

Keywords Structural stresses • Hole-drilling • DIC • Stress datum • Residual stress

11.1 Introduction

The ability to measure internal loads within engineered structures is important because it provides information critical to evaluating structural safety and maintenance needs [1, 2]. It also provides a means to monitor changing loads, support subsidence and structural integrity. The challenge when seeking to measure the stresses that exist within a structure is in establishing a zero stress reference datum. This is further complicated by the presence of redundancies in a structure that allow internal stresses to exist without the application of any external load. From a measurement standpoint the stresses within a structure can be divided into two main types: (1) relative measurements that do not require a zero stress datum, and (2) absolute measurements that do. Evaluations of the stress due to a vehicle on a bridge or from the wind on a building are examples of relative measurements because they are done by observing the changes caused by the presence and absence of these loads. More difficult to measure are stresses due to gravity, residual stresses, or stresses due to structural redundancy because such loads are not readily removed so that the changes caused by their presence and absence cannot be directly observed. Different loading types will affect structures in different ways, so for a measurement method to be useful, it must be able to identify and isolate all stresses that can exist within a structure.

A typical way to evaluate stress is to use a strain gauge to measure surface strain as a load is applied or removed [3]. The associated stress is then inferred from the measured strain. This procedure is useful for determining Type (1) relative stresses because it demands an ability to apply and remove the associated load. However, this approach is generally insufficient for the measurement of Type (2) absolute stresses. Alternative absolute methods have been developed to relate specific material properties to the internal stress state of a material. For example, magnetic properties involving Barkhausen noise and the Villari effect [4], can be related to the stress state in a material, but they require careful calibration with respect to a material's mechanical and metallurgical properties. Crystal lattice strain, measured using X-ray diffraction [5], can be used to determine the state of stress in a material but is limited to very shallow depths, typically just a few microns. Alternatively,

J. Harrington • G.S. Schajer (✉)
Department of Mechanical Engineering, University of British Columbia, Vancouver, Canada
e-mail: schajer@mech.ubc.ca

Y. Zhu, A.T. Zehnder (eds.), *Experimental and Applied Mechanics, Volume 4*,
Conference Proceedings of the Society for Experimental Mechanics Series, DOI 10.1007/978-3-319-42028-8_11

the ultrasonic wave velocity in a material can be correlated to internal stress [4], but to have an accurate measurement, the material must cycle though the zero stress datum, which is not always possible. Destructive stress measurement methods, such as the sectioning, slitting and contour methods [6] are effective, but the damage that they create is so large as to render the measured material no longer functional. Thus, they are not useful for the measurement of structural stresses.

A "semi-destructive" approach, the hole-drilling method, was originally developed for the measurement of residual stresses [7, 8]. This method involves drilling a small hole in the test object. The removal of the stressed material originally within the hole causes the stresses in the surrounding material to redistribute. The resulting deformations around the hole can be measured and correlated to the residual stresses originally within the removed material. This approach is of interest here because the free surface of the hole provides the desired zero datum for measurement of the Type (2) absolute stresses. The drilled hole is of modest size and is generally sufficiently small that its presence does not adversely affect the functionality of the structure.

In the work described here, it is proposed to apply and adapt the hole drilling method to enable it to be used for field measurements of Type (2) absolute stresses in engineered structures. This is done by enlarging the typical scale of the hole-drilling technique to that of common engineered structures. To simplify operation under field conditions, the use of Digital Image Correlation (DIC) [9, 10] is explored as a means of measuring the deformations due to drilling. The full-field measurements provided by DIC also enable mathematical compensations to be made for environmental effects such as temperature changes [11]. To explore the applicability of the proposed method, the measurement of four different structure types will be examined.

11.2 Background

11.2.1 Digital Image Correlation

Digital Image Correlation is a popular and highly developed metrology that utilizes image processing techniques to determine the displacements within a region of interest (ROI) shared by two images, one taken before (reference) and one taken after (deformed) a deformation [9]. Figure 11.1 shows the setup typically used for a 2D single-camera DIC measurement.

The specimen either has a natural surface texture or an applied random speckle pattern painted on. This surface pattern is imaged by the camera before and after the deformation of interest. The analysis procedure involves dividing the initial image into small regions called patches and then using correlation techniques to locate the corresponding regions in the second image [9]. The relative positions of the patches indicate the material displacements between the two images. The combined results from all patches give the displacement fields in the X and Y directions throughout the target ROI. Under good conditions, the displacements can be determined to within 0.02 pixels or better [11]. In the present work, the ROI is the area of material surrounding the hole drilled for hole-drilling residual stress measurements.

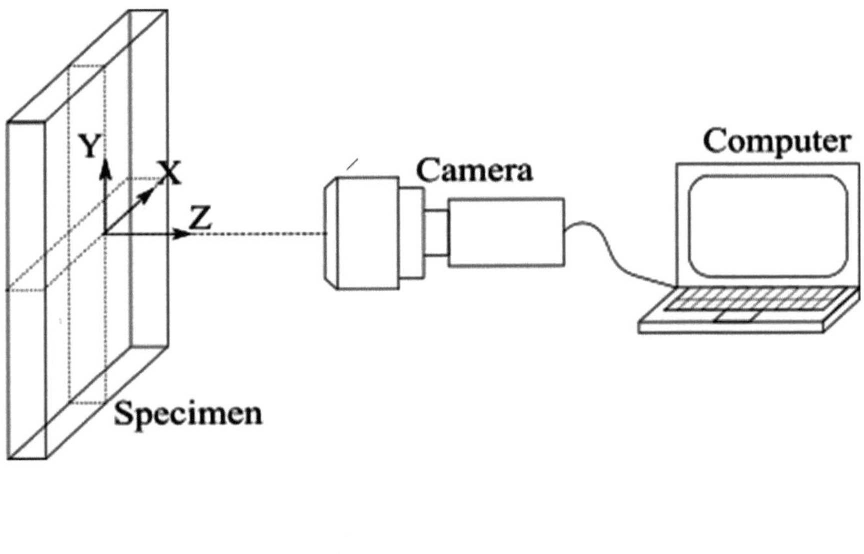

Fig. 11.1 Typical 2D DIC measurement setup

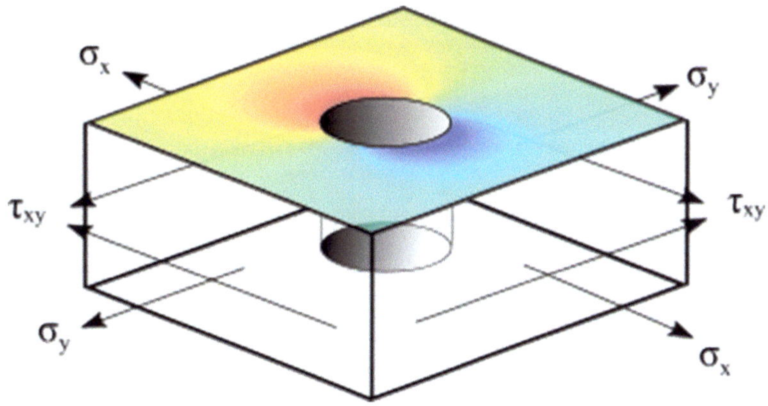

Fig. 11.2 Hole-drilling measurement geometry

11.2.2 Hole-Drilling Method

Figure 11.2 illustrates the hole-drilling method [12]. A hole is drilled in a stressed material, as shown. The drilling of the hole and the consequent release of the stresses σ_x, σ_y, and τ_{xy} originally contained within the hole causes a redistribution of the stresses in the remaining material around the hole. By measuring the deformation of the material around the hole at the surface (shown in color), the stresses originally present in the material within hole can be mathematically evaluated [12]. This measurement is commonly done using strain gauges [13, 14], but in recent years full-field optical techniques such as ESPI [15], Moiré interferometry [8] and DIC [2, 11, 16] have been used. Here, the DIC approach is chosen because it is robust to use in field conditions and it avoids the substantial preparation and installation time associated with strain gauges.

The full-field optical measurements of the deformations around a drilled hole have an advantageous feature over the traditional strain gauge technique in that they have a data content much greater than the minimum requirement, with one or more measurements at each image pixel. These excess data are useful for noise averaging and also for the identification and removal of systematic artifacts. This approach has been used previously for artifact identification and correction with ESPI data and DIC data [11, 17]. The technique described in [11] will be used here for artifact removal.

11.3 Method Development

11.3.1 Annular Cutter Evaluation

The ability of the hole-drilling method to identify interior stresses is limited to a depth of one hole radius [8]. Typical hole drilling measurements done according to ASTM E837 [12] are made using holes up to about 2 mm in diameter, allowing for stresses to be measured up to a depth of about 1 mm. To evaluate the characteristic stresses within a structure, it is necessary to reach into at least one half the material thickness. Depending on the component size, this distance can range from several millimeters to several centimeters. This means that the hole used for the measurement of structural stresses needs to be significantly larger than is traditionally used.

The DIC approach enables surface displacement measurements to be made at any needed scale simply by adjusting the imaging magnification. This means that custom gauges or setups are not required to make structural stress measurements over a wide range of material thicknesses.

Annular cutters of the type shown in Fig. 11.3 were chosen to allow the hole-drilling method to be adapted for field use. Such cutters are commonly used in the fabrication and repair of steel structures, and are known for their effectiveness and ease of use [18].

The use of DIC for hole-drilling measurements is relatively new, so to investigate any artifacts associated with the use of large annular cutters, an initial study using Electronic Speckle Pattern Interferometry (ESPI) was done. The ESPI approach was effective for these test measurements in the laboratory, but for field use DIC is preferred because of its much greater ruggedness.

Figure 11.4 shows the setup used for annular cutter evaluation. The objective was to explore various cutting procedures so as to produce a clean cut without significant introduction of machining stresses. In addition, it was important to control the

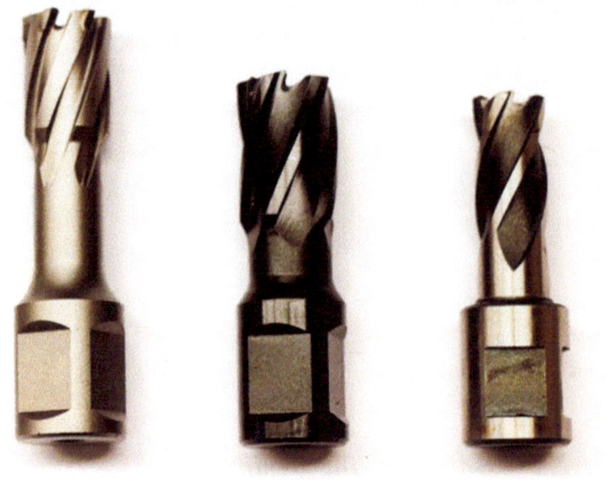

Fig. 11.3 Example annular cutters

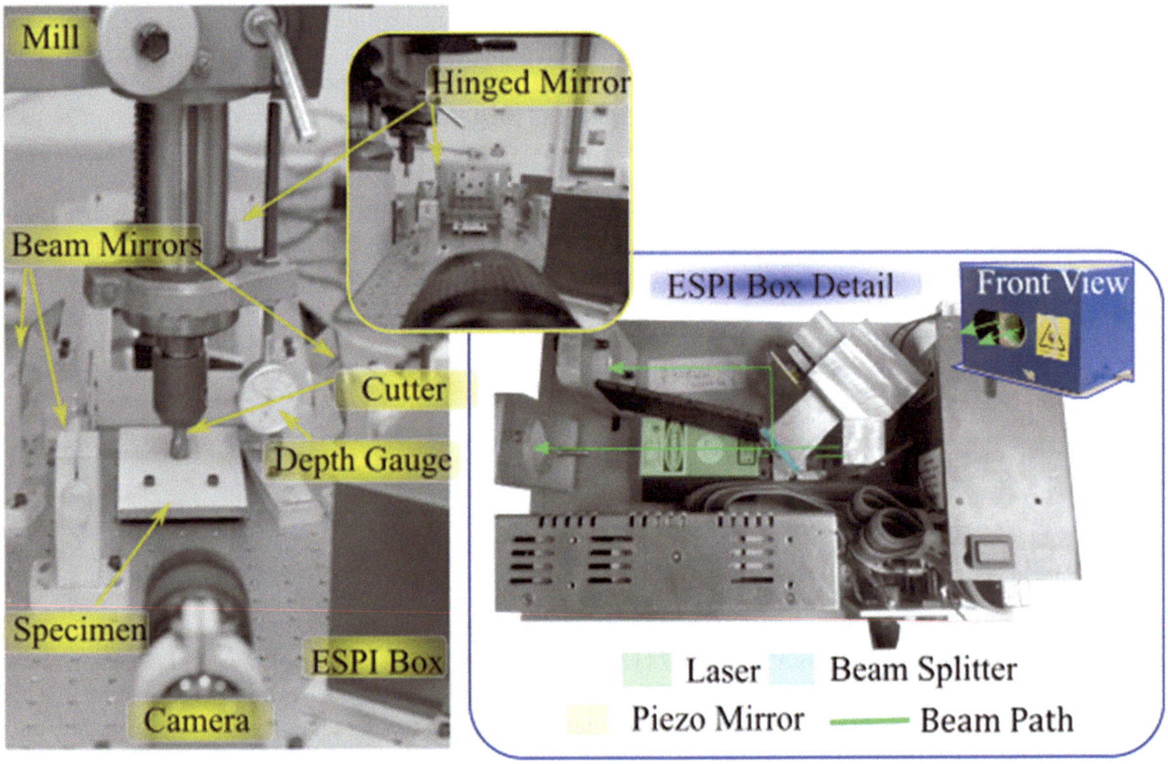

Fig. 11.4 Annular cutter evaluation interferometry setup

chip formation to produce many small chips because long curling chips tended to spin round and scratch the measurement surface. This chhip length reduction was done by cutting with a slowly pulsing feed rate.

The possible creation of machining stresses during annular hole drilling was investigated by doing cutting tests on stress-free steel plates. Specimens of 1080 hot rolled steel plate were prepared by soaking at 650 °C for 24 h followed by slow cooling. Any stresses introduced by the hole drilling would then be observable in the residual stress measurements. Initial tests indicated that the drilling process introduces significant heat to the area around the hole, causing surface displacements much larger than those due to machining stresses. This artifact has the form of an equal stretch in the x and y directions and can be removed mathematically using an artifact correction scheme previously developed for hole-drilling measurements [11, 15]. Figure 11.5 demonstrates this ability of remove the displacements due to thermal expansion. The top row of ESPI fringe patterns were measured, representing the x-direction displacements at 1, 3, 5, and 9 min after drilling. The second row

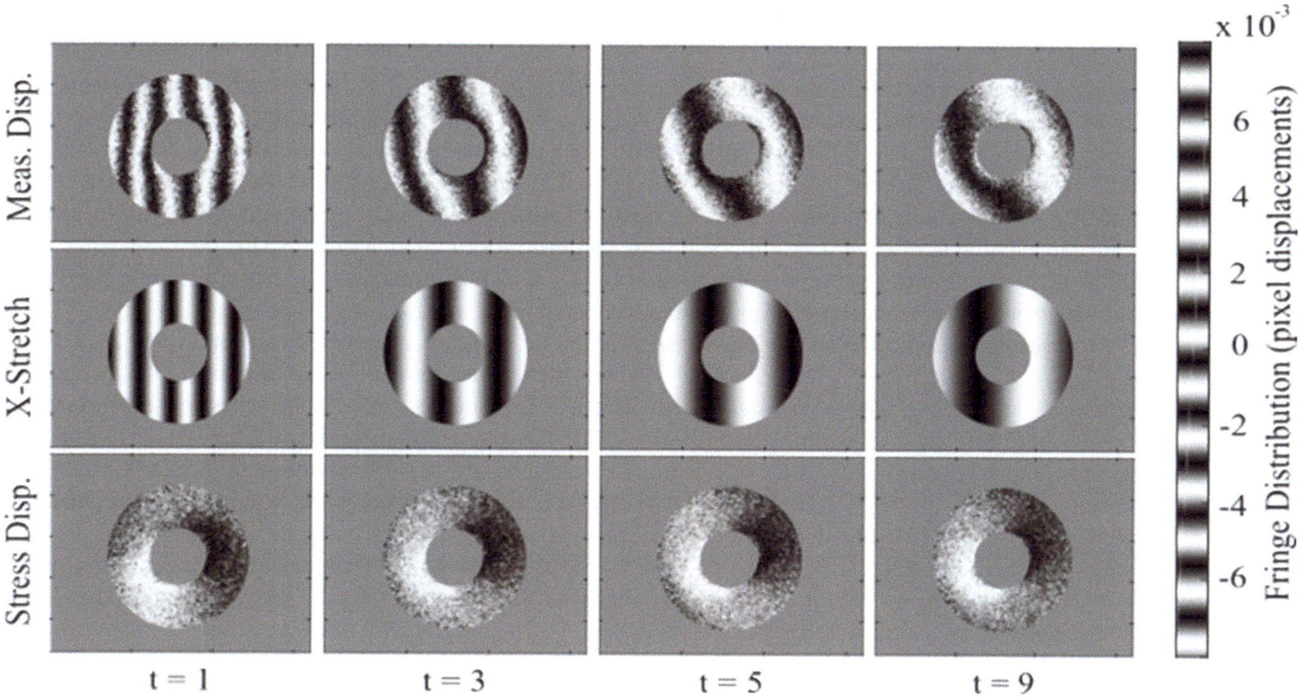

Fig. 11.5 Correction of thermal effects at times t = 1, 3, 5 and 9 min after hole drilling. *First row*: raw measured ESPI fringes, *second row*: mathematically estimated thermal artifacts, *third row*: drilling effects observed by subtracting second row from first row

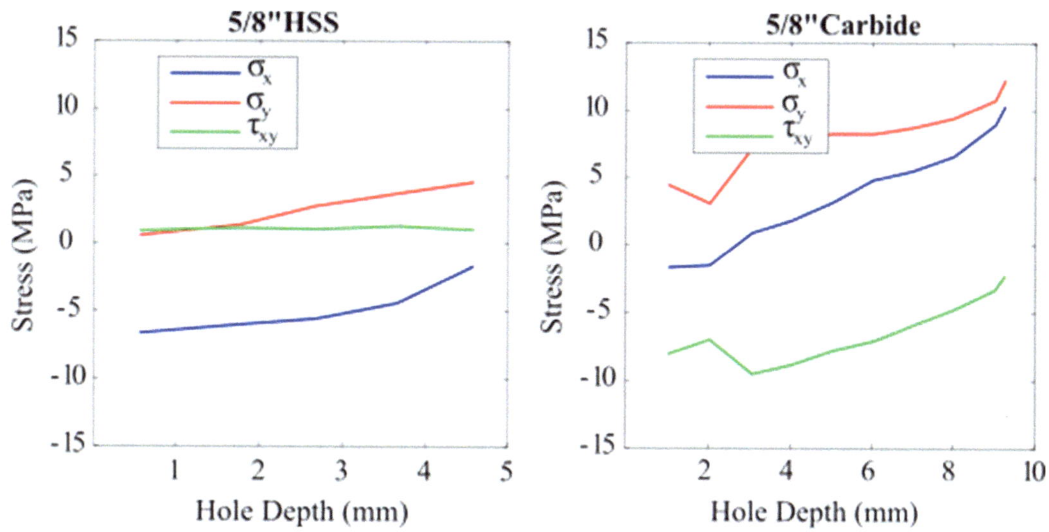

Fig. 11.6 Cutting stress profiles observed in stress relieved 1080 plate steel with high-speed steel (*left plot*) and carbide (*right plot*) 5/8″ annular cutters

shows the mathematically estimated stretch artifact due to heating, and the third row shows the net results after subtracting the estimated artifacts from the raw measurements on the first row. It can be seen that this process works very effectively and similar corrected fringe patterns are observed in all cases.

Effective removal of the thermal artifact allowed direct evaluation of the machining induced stresses from the annular cutter. Figure 11.6 shows results from tests on 5/8 in. carbide and 5/8 in. high-speed steel annular cutters. These measured stresses are consistently less than 10 MPa, which is within the resolution of a typical hole drilling measurement. Thus, the annular cutting process can be considered to be effectively "stress-free".

11.4 DIC Measurements

The initial tests using ESPI gave confidence in the use of annular cutters for hole drilling. Subsequent measurements were made using DIC, with the experimental setup shown in Fig. 11.7. For the DIC measurements, the specimens were lightly painted white around the hole area and then distantly sprayed with black paint to create a random pattern of black speckles to be used for the DIC measurements. Finally, a protective coating of polyurethane was applied so that any metal dust created during cutting could be wiped away easily. Importantly, the coating facilitates the use of cutting oil during drilling, which should further reduce machining stresses. Excess oil could be wiped off prior to the imaging for DIC measurements.

The setup in Fig. 11.7 was arranged within a 300 kN capacity Tinius Olsen compression testing machine. Structural specimens ~40 cm long were set in the machine. Strain gauges were attached to the front and back faces of the specimens so the uniformity of the load application could be monitored and adjusted if necessary. A camera was mounted adjacent to the specimen to image the speckle pattern around the drilled hole and provide the data for the DIC calculations, which were done using NCOOR software [19]. Within this physical arrangement it was not practical to drill the holes in-situ. Instead the specimens were removed to be drilled using a nearby drill-press and then returned their position within the test setup.

Four structural elements were tested: a channel, an I-beam, a square tube and a 136 lb mainline railway rail. Table 11.1 details the dimensions of these specimens. Tests were done by imaging the hole location starting at zero applied load and then at several increments of applied load. By comparing each image to the image taken with no compressive loading the surface displacements due to each compressive force can be measured and the stresses calculated accordingly. By using the zero load image as reference it allows for all other stresses that may exist in the material, such as residual stresses or machining induced stresses, to be subtracted from the calculation.

Figure 11.8 shows a typical surface displacement measurement when using DIC. The displacements are presented in "fringe" format to allow easy comparison with the prior ESPI measurements. The fringes correspond to contours of equal surface displacement, similar to the height contours on a topographic map. This format is useful because it is able to show both large and small displacements within one scale. The figure illustrates the capability of DIC to capture the displacements

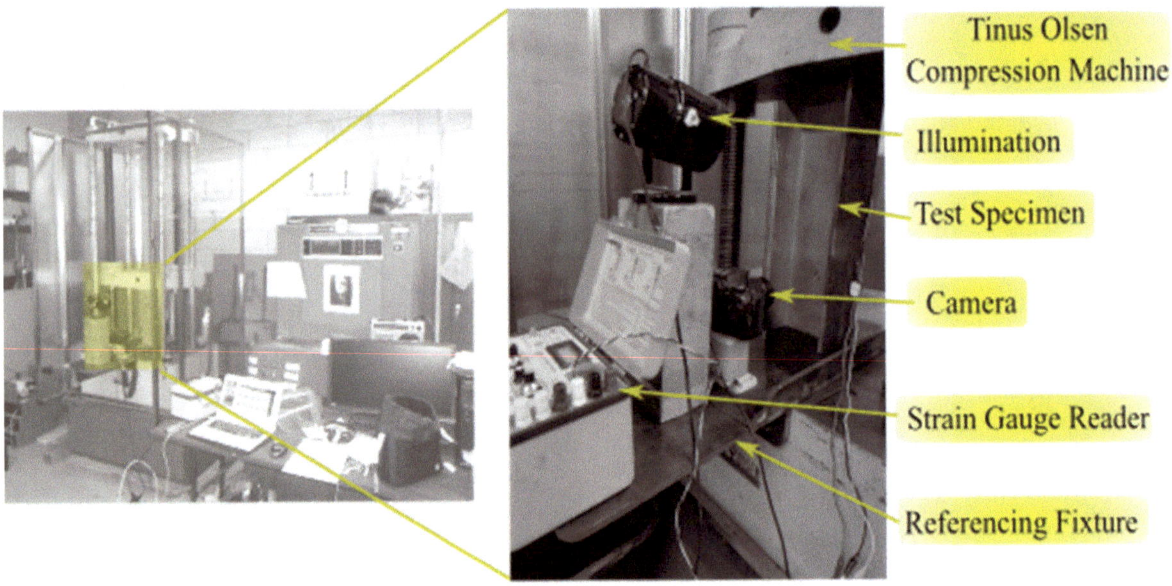

Fig. 11.7 Experimental setup for DIC/ hole-drilling measurements with applied loading

Table 11.1 Measurement validation and accuracy experiment details

	Channel	I-beam	Square tube	Rail
Hole radius (mm)	6.35	8.73	8.73	7.94
Material thickness (mm)	5.25	6.75	6.25	17.45
Hole depth (mm)	4.05	3.50	3.00	4.25
Pixel density (pixel/mm)	51.25	59.82	68.24	61.55
Stress range (MPa)	0–150	0–75	0–100	0–30

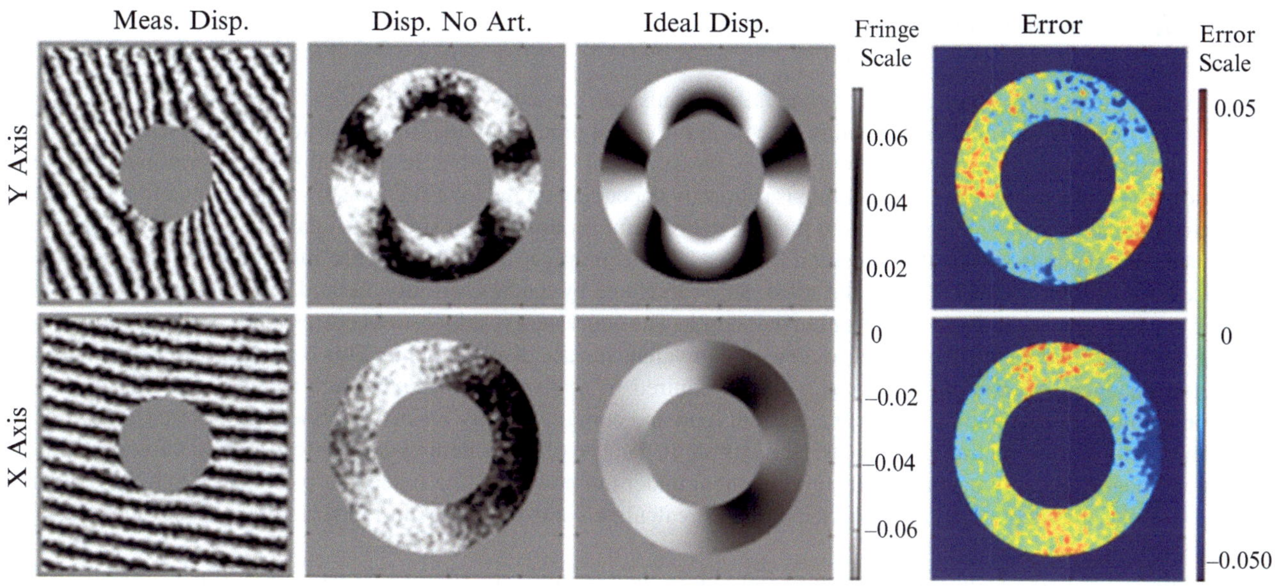

Fig. 11.8 Example of measured surface displacements on I-beam loaded to 75 MPa. *First row*: y-displacements, *second row*: x-displacements. *First column*: measured displacements, one fringe corresponds to 2.5 μm, *second column*: displacements after removal of thermal artifacts, *third column*: "ideal" displacements corresponding to the applied stresses, *fourth column*: residual errors between second and third column displacements

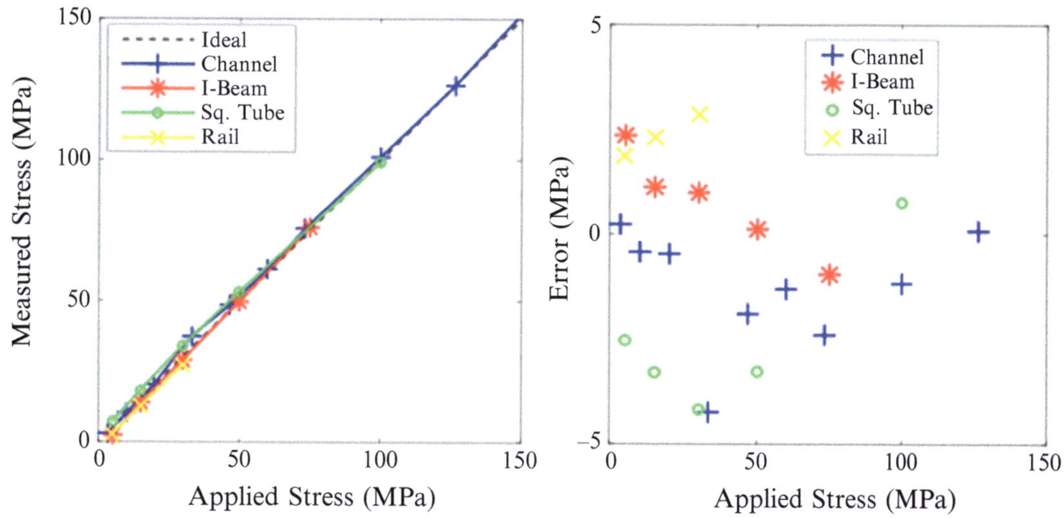

Fig. 11.9 Stress measurement results for the measurements described in Table 11.1 showing the measured stress and error relative to the applied stress

resulting from hole-drilling as well as the robustness of the artifact correction method and its ability to isolate displacements due only to the redistribution of stresses. The first column shows that there is some specimen rotation between measurements. Even with this effect, the displacements with the artifacts mathematically removed match quite closely to the ideal displacements. Not shown in this example measurement are displacements due to rigid body motion between images, which in some cases could be greater than 30 pixels. These rigid-body motions are mainly due to small errors in replacing the specimens in position after removal for hole-drilling.

Figure 11.9 shows the longitudinal stress values calculated from the DIC measurements for each structural member and also the associated measurement error over a range of applied stresses. The measured stress values closely follow the linear trend of the ideal measurement and that the deviations lie within $\pm$ 5 MPa. Based on these experiments, it is clear that even with imperfect displacement measurements, the stress calculation method is capable of measuring the applied loads reliably.

11.4.1 Residual Stresses

Elements used in structures, such as I-beams and railroad rails, can contain significant residual stresses due to manufacturing processes, particularly the roller straightening process at the end of production [6, 20]. For structural analyses, it important to separate the residual stress component from a stress measurement. For example, in the evaluation of continuously welded rail, structural failure such as thermal buckling is primarily dependent on the longitudinal stresses due to temperature change and not to residual stress. For a measurement method to be useful for this type of structural analysis, residual stresses need to be isolated and separated from the structural stresses. To accomplish this, the concept of equivalent residual stress is introduced. Equivalent residual stress is the stress state calculated by the hole drilling method at a particular hole depth due to all the residual stresses in the material within that depth. The equivalent residual stress corresponds to the size of a uniform residual stress within the hole depth that would give equivalent surface displacements. This equivalent stress changes as the hole depth changes. A calibration measurement is required to apply this concept to actual measurements for a given structural element. A specimen without any external load and containing only residual stresses is measured at several hole depths through the cross-section of the material resulting in a curve of equivalent residual stress as a function of depth. In a subsequent measurement of a similar structural element under unknown loading made at a particular depth, the equivalent residual stress at that depth can be subtracted from the calculated stresses to obtain the structural stresses independent of residual stress.

Using the same experiment setup shown in Fig. 11.7, eight additional experiments were performed, four to establish equivalent residual stress curves and four additional measurements made at various loading conditions. With these measurements, rather than using a zero-load reference image, a zero-depth reference image is used. In consequence, the resulting stress measurements will contain all applied, machining-induced and residual stresses. The equivalent residual stress curves can then be subtracted from the calculated stress values to find the applied stresses. Note, that for the experiments discussed, stresses are only calculated in the longitudinal direction. Table 11.2 lists the details for these experiments.

Figure 11.10 shows the equivalent residual stress calibration curves for the four structural element types. These measurements show a big variation in the scale of the residual stresses. Without a method to determine the equivalent residual stress calibration curves, it is impossible to determine the applied structural stress.

Figure 11.11 shows the corrected structural stress results and associated error. These results show the measurement correction method with each of the four structure types and that the resulting error is only slightly enlarged beyond that of the pervious measurements, at ±6 MPa. The increased measurement error is likely due to the variation in the machining stresses between the measurements used to create the calibration curves and the subsequent corrected measurements. Successfully measuring the stresses applied to each structural member, on separate specimens from those used to create the calibration curves, shows that the consistency of the stress measurements and that the same stresses can be calculated independent of the cutter used. This demonstrates that the developed DIC/hole-drilling stress measurement method is a practical choice to measure structural stresses over a broad range of structure types.

Table 11.2 Equivalent residual calibration and corrected measurements

	Channel	I-Beam	Square tube	Rail
Equivalent residual stress calibration measurements				
Hole radius (mm)	6.35	8.73	8.73	7.94
Material thickness (mm)	5.25	5.91	6.15	17.45
Hole depth (mm)	2.14–5.25	2.75–5.91	2.00–6.15	1.75–9.75
Pixel density (pixel/mm)	51.25	59.82	68.24	61.55
Corrected stress experiments				
Hole radius (mm)	9.53	7.94	7.94	7.94
Material thickness (mm)	5.25	5.91	6.15	17.45
Hole depth (mm)	4.00	3.25	3.25	4.00
Pixel density (pixel/mm)	51.25	59.82	68.24	61.55
Applied Stress (MPa)	0–100	0–50	0–100	0–30

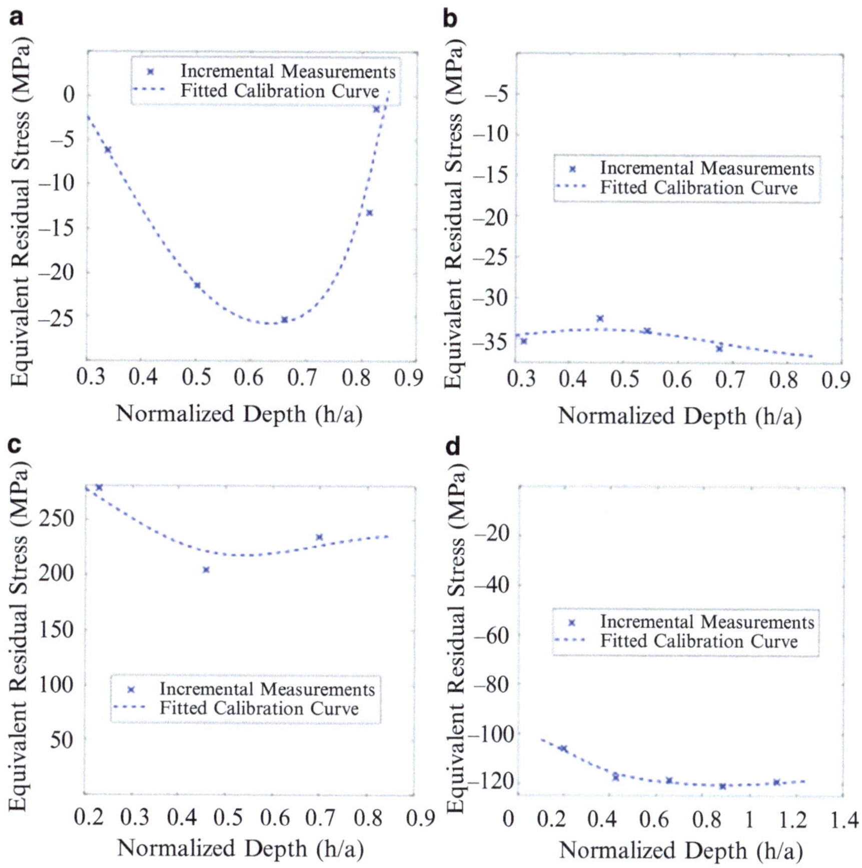

Fig. 11.10 Equivalent residual stress calibration curves for channel (**a**), I-Beam (**b**), square (**c**), and rail (**d**) structural elements showing the equivalent residual stress as a function of normalized hole depth. Normalized hole depth is hole depth divided by hole radius

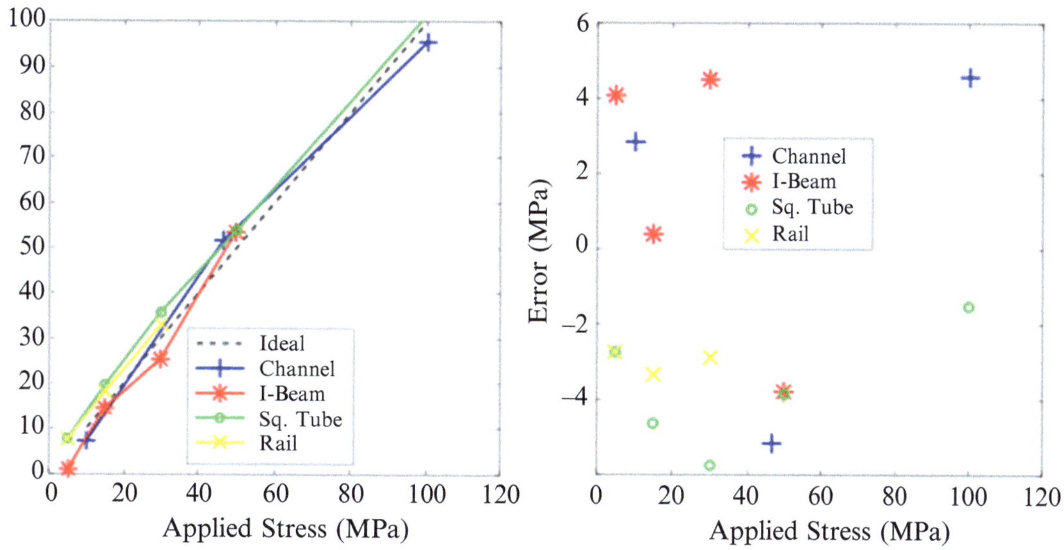

Fig. 11.11 Corrected Stress measurement results for the measurements described in Table 11.2 showing the correct stress and error relative to the applied stress

11.5 Conclusions

This work was designed to develop and demonstrate a practical method to measure structural stresses that may be used in industry. For effective use, the method must be repeatable, accurate and quick, even with disturbances present in an industrial setting. To realize this goal, a measurement method combining the hole-drilling technique and digital image correlation was developed.

Initial ESPI test measurements demonstrated the effectiveness of annular cutters for hole-drilling measurements. An accuracy of ~±5 MPa was achieved. Annular cutters are already widely used in the field for structure fabrication, so their adoption of this measurement method is a practical choice.

Subsequent experiments conducted to evaluate the effectiveness of DIC/hole-drilling measurements on four typical structure types showed that DIC/hole-drilling measurements can measure stresses to within ±6 MPa. This was the case even with significant artifacts present in the calculated DIC displacements, as would also arise in field measurements.

Finally, a method was developed and evaluated for the subtraction of residual stress components from a given measurement. The developed method uses a calibration measurement to determine the equivalent residual stress profile that would exist for a given structural element type. The equivalent residual stress could then be subtracted from the hole-drilling measurement to give the desired structural loads. The main drawback of this method is that it requires a calibration measurement. Future work is anticipated to estimate the equivalent residual stress using FEA models to determine equivalent stress.

Acknowledgments This work was financially supported by the Natural Science and Engineering Research Council of Canada (NSERC) and by American Stress Technologies, Inc, Pittsburgh, PA, USA. Dr. Richard Stock, LB Foster Rail Technologies Corp., Burnaby, Canada kindly provided rail specimens.

References

1. Kish, A., Samavedam, G.: Track Buckling Prevention: Theory, Safety Concepts, and Applications. DOT/FRA/ORD-13/16, Federal Railroad Administration, Washington, DC (2013)
2. McGinnis, M.J.: Application of three-dimensional digital image correlation to the core-drilling method. Exp. Mech. **45**(4), 359–367 (2005)
3. Dally, J.W., Riley, W.F.: Experimental Stress Analysis. College House Enterprises, LLC, Knoxville (2005)
4. Koob, M.J.: The development of a vibration technique for estimation of neutral temperature in continuously welded railroad rail, MS thesis, Civil Engineering, University of Illinois at Urbana-Champaign, (2005)
5. Fitzpatrick, M.E., Fry, A.T., Holdway, P., Kandil, F.A., Shackleton, J., Suominen, L.: Determination of Residual Stresses X-ray Diffraction—Issue 2. Measurement Good Practice Guide. National Physical Laboratory, Teddington (2005)
6. Schajer, G.S. (ed.): Practical Residual Stress Measurement Methods. Wiley, Chichester (2013)
7. Mathar, J.: Determination of inherent stresses by measuring deformations of drilled holes. Trans. Aerican Soc. Mech. Eng. **56**(2), 249–254 (1934)
8. Schajer, G.S.: Hole-drilling residual stress measurements at 75: origins, advances, opportunities. Exp. Mech. **50**(2), 245–253 (2010)
9. Pan, B., Qian, K., Xie, H., Asundi, A.: Two-dimensional digital image correlation for in-plane displacement and strain measurement: a review. Meas. Sci. Technol. **20**(6) (2009)
10. Sutton, M.J., Orteu, J.-J., Schreier, H.W.: Image Correlation for Shape, Motion and Deformation Measurements. Springer, New York (2009)
11. Schajer, G.S., Winiarski, B., Withers, P.J.: Hole-drilling residual stress measurement with artifact correction using full-field DIC. Exp. Mech. **53**(2), 255–265 (2012)
12. ASTM.: Determining Residual Stresses by the Hole-Drilling Strain-Gage Method, ASTM Standard Test Method E837-13. American Society for Testing and Materials, West Conshohocken (2013)
13. Schajer, G.S.: Measurement of non-uniform residual stresses using the hole-drilling method. Part I—stress calculation procedures. J. Eng. Mater. Technol. **110**(4), 338–343 (1988)
14. Schajer, G.S.: Measurement of non-uniform residual stresses using the hole-drilling method. Part II—practical application of the integral method. J. Eng. Mater. Technol. **110**(4), 344–349 (1988)
15. Schajer, G.S.: Full-field calculation of hole drilling residual stresses from electronic speckle pattern interferometry data. Exp. Mech. **45**(6), 526–532 (2005)
16. Baldi, A.: Residual stress measurement using hole drilling and integrated digital image correlation techniques. Exp. Mech. **54**(3), 379–391 (2013)
17. Ponslet, E., Steinzig, M.: Residual stress measurement using the hole drilling method and laser speckle interferometry part II: analysis technique. Exp. Tech. **27**(5), 45–48 (2003)
18. Stephenson, D.A., Agapiou, J.S.: Metal Cutting Theory and Practice. CRC Press, New York (2005)
19. Blaber, J., Adair, B., Antoniou, A.: Ncorr: open-source 2D digital image correlation Matlab Software. Exp. Mech. **55**(6),1105–1122 (2015)
20. Schleinzer, G., Fischer, F.: Residual stress formation during the roller straightening of railway rails. Int. J. Mech. Sci. **43**(10), 2281–2295 (2001)

Chapter 12
A Simulator to Optimize the Experimental Set-Up for Elasto-Plastic Material Characterization

Michele Badaloni, Attilio Lattanzi, Marco Rossi, Pascal Lava, and Dimitri Debruyne

Abstract In the field of material characterization, the use of Digital Image Correlation (DIC) is coupled with an inverse methodology, such as the Virtual Fields Method (VFM), is getting an increasing interest in the recent years. That allows identifying material properties by performing tests on specimens with heterogeneous stress-strain fields. Respects to standard techniques, VFM consents to reduce the number of tests required to identify complex constitutive models. However, that methodology is sensitive to the adopted experimental set-up. The specimen geometry and the parameters used in the DIC settings can influence remarkably the identification results. Therefore, a preliminary study is necessary to determine the best experimental set-up to apply. To this purpose a simulating procedure, which reproduces numerically the whole measurement chain, was developed. The simulator includes most of the experimental uncertainties typically present in DIC measurements, moreover is able to look at different specimen geometries. The simulator was used here to optimize the geometry of specimens to be used to identify the elasto-plastic behavior of isotropic materials.

Keywords Simulated experiments • Virtual fields method • Inverse identification • Digital image correlation • Large strains

12.1 Introduction

Nowadays, the strengthened usage of Finite Element Analysis in both academic and industrial practice requires a correct calibration of the constitutive model and corresponding parameters. To this purpose, several typologies of experimental tests have been developed according to the needed material's property. In this scenario a crucial role is played, obviously, by the measurement techniques, which represents an essential tool to retrieve information on materials' performances. From this standpoint, the full-field measurement permits to obtain a larger number of information compared to standard laboratories techniques, and it can be applied to study, for instance, heterogeneous behaviour like anisotropy, localized phenomena, etc. [1, 2]. The Digital Image Correlation (DIC) [3] is one of the most widespread full-field techniques, thanks to its simple set-up arrangement and relatively low-cost requirements, and represents a powerful instrument if coupled with an inverse method like the Virtual Fields Method (VFM) [4] in order to identify the constitutive parameters.

The exactness of the identification process through VFM is deeply influenced by the accuracy of displacement field measurement on the specimen's surface obtained by the full-field technique. In fact, as widely reported in literature [5–7], factors such spatial resolution, noise intensity and quality of the speckle pattern strongly affect the identification results, acting as error sources whose weights are not still completely understood. In order to face this issue, Rossi and Pierron [8] developed a numerical procedure to simulate the whole measurement chain in the case of the unnotched Iosipescu test, using the Grid Method technique [9] to obtain the strain field. The aforementioned procedure was also applied to DIC [10]

M. Badaloni (✉)
Università Politecnica delle Marche, via Brecce Bianche, Ancona 60100, Italy

Department of Materials Engineering, K.U. Leuven, Campus Gent, Gebroeders De Smetstraat 1, Ghent 9000, Belgium
e-mail: m.badaloni@univpm.it

A. Lattanzi • M. Rossi
Università Politecnica delle Marche, via Brecce Bianche, Ancona 60100, Italy
e-mail: a.lattanzi@pm.univpm.it; m.rossi@univpm.it

P. Lava • D. Debruyne
Department of Materials Engineering, K.U. Leuven, Campus Gent, Gebroeders De Smetstraat 1, Ghent 9000, Belgium
e-mail: pascal.lava@kuleuven.be; dimitri.debruyne@kuleuven.be

Y. Zhu, A.T. Zehnder (eds.), *Experimental and Applied Mechanics, Volume 4*,
Conference Proceedings of the Society for Experimental Mechanics Series, DOI 10.1007/978-3-319-42028-8_12

and used to design optimized DIC set-up configurations, for instance in case of polymeric foams [11]. Moreover, Badaloni et al. [12] employed the simulator as a diagnostic tool to investigate the impact on parameter identification of different experimental uncertainties, as out-of-plane and in-plane movements respectively, noise and illumination conditions.

So far, the simulator was mainly applied to elasticity while the VFM can be also applied to several non-linear cases. In particular, plasticity represents one of the most attractive field [13], many applications can be found in the literature, for example in large strain plasticity [14], in case of anisotropic behaviour with models as Hill48 [15] and YLD2000-2d [16], and to evaluate post-necking regime [17]. However, for all those cases, the definition of the optimal specimen geometry is still debatable.

The present paper describes a numerical method that exploits the simulator and the VFM to optimized the geometry of specimens that will be used for characterizing the hardening behaviour of metals at large strains. In particular, it focuses on effects of notches and experimental uncertainties. Albeit the present approach is limited to isotropic plasticity, it can be easily extended to more complex constitutive models [18, 19].

12.2 Optimization Process

The developed optimization process algorithm basically can be divided in two subsequent phases, as outlined in Fig. 12.1: the simulated experiment, that produces synthetic images required to the second phase, the identification procedure. The main input is the geometry of the specimen, whose features are the objects of optimization. The generated geometry is submitted to Finite Element analysis. Since isotropic material was taken into consideration, its behaviour is determined only by a reference hardening curve, expressed, in this case, by the Swift's law:

$$\sigma_{eq} = K(\varepsilon_0 + \overline{\varepsilon})^N \tag{12.1}$$

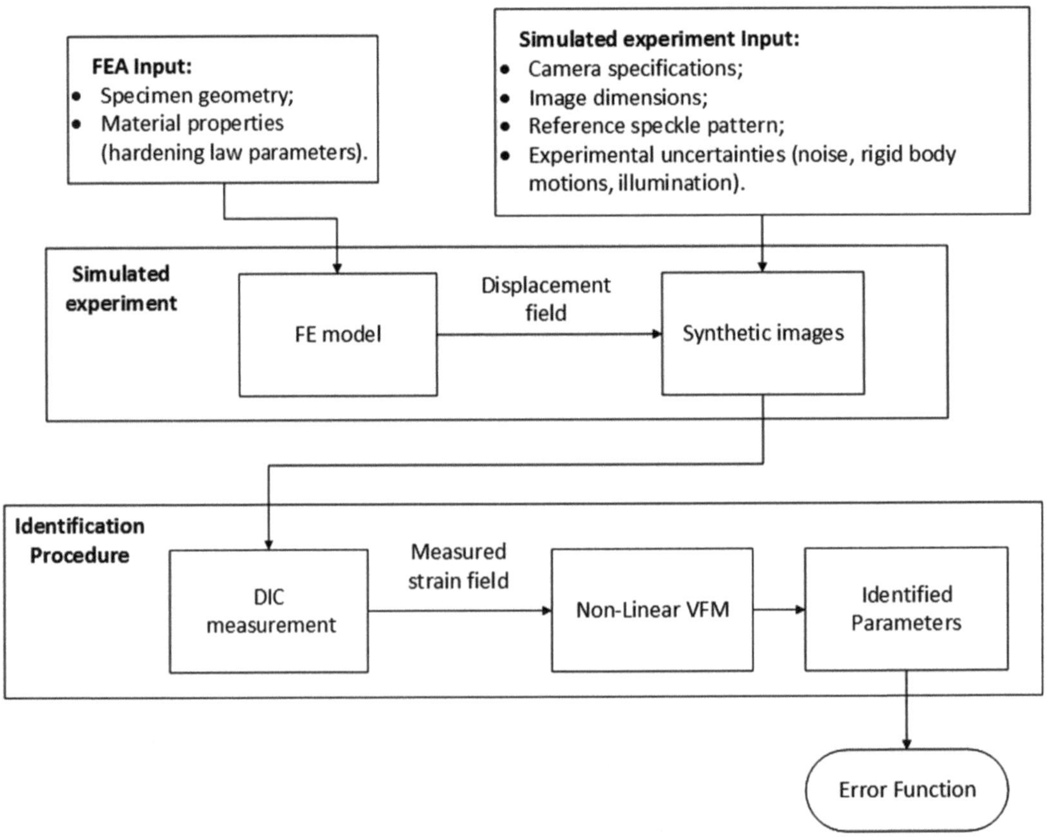

Fig. 12.1 Flowchart of the optimization process

The FEA constitutes the first step to create the synthetic images. A reference image is generated starting from nodal coordinates of the numerical model, the dimensions (expressed in pixels) of camera sensor and a reference speckle pattern; then, the deformed images are constructed deforming this reference images according to the displacement vector at each pixel location calculated with shape functions of the corresponding element. In order to evaluate the impacts on the identification of error sources typical of DIC measurements, experimental uncertainties are introduced in the simulated images to keep into account noise, rigid body movements and illumination variations.

DIC is the applied to the synthetic images to obtain the displacement and the strain field, using the same method applied in the actual case.

The identification is achieved by an inverse method, the VFM. This methodology is based on the Principle of virtual work and allows to determine the material constitutive parameters from strain field measured on specimen's surface. In non-linear cases such as plasticity, the constitutive equations do not depend linearly from these parameters, which become solution of a set of non-linear equations solved, generally minimizing a certain cost function [4]. Therefore, indicating all parameters that regulates the material behaviour with the vector $\boldsymbol{\xi}$, the non-linear VFM is applied using the approach widely described in [14, 16], varying $\boldsymbol{\xi}$ in order to minimize the cost function $\Psi(\boldsymbol{\xi})$:

$$\Psi(\boldsymbol{\xi}) = \sum_{i=1}^{N_{vf}} \sum_{j=1}^{N_{step}} \left| \int_{\Omega} \mathbf{T}_j^{1PK} : \delta\mathbf{F}_i^* dV - \int_{\delta\Omega} \left(\mathbf{T}_j^{1PK} \mathbf{n}_0 \right) \cdot \delta v_i^* dS \right|. \tag{12.2}$$

In Eq. (12.2) $\mathbf{T}^{1PK}$ indicates the first di Piola-Kirchhoff tensor, $\boldsymbol{\delta}\mathbf{v}^*$ any kinematically admissible virtual displacement field, $\boldsymbol{\delta}\mathbf{F}^*$ the corresponding virtual displacement gradient tensor, Ω the solid's volume, $\partial\Omega$ is the boundary surface and $\mathbf{n_0}$ the unit vector normal to the surface in the undeformed configuration. Two terms can be distinguished in Eq. (12.2): the former represents the internal virtual work, while the latter is virtual work made by external forces; in equilibrium conditions this difference is equal to zero. The cost function $\Psi(\boldsymbol{\xi})$ is evaluated for each N_{vf} virtual fields chosen by user and each N_{step} measurement step.

Finally, the goodness of identification results is assessed with the following error function:

$$RMSE = \sqrt{\frac{1}{\overline{\varepsilon}_{\lim}} \int_0^{\overline{\varepsilon}_{\lim}} \left(\overline{\sigma}_{ref} - \overline{\sigma}_{VFM} \right)^2}, \tag{12.3}$$

which indicate the Root Mean Square Error of the reference and identified stress-strain curve respectively, due to the fact that different parameters have different impacts in the description of hardening curve. $\overline{\varepsilon}_{\lim}$ represents the maximum equivalent plastic strain considered.

12.3 Simulated Experiment

In this study the shape of the specimen is characterized by the presence of two notches, whose geometry is regulated by seven independent variables (h, x_L, x_R, y_L, y_R, θ_L, θ_R), as displayed in Fig. 12.2. In particular, 50 different configurations are composed on the basis of authors' experience, and all the corresponding variables' values are reported in Table 12.1.

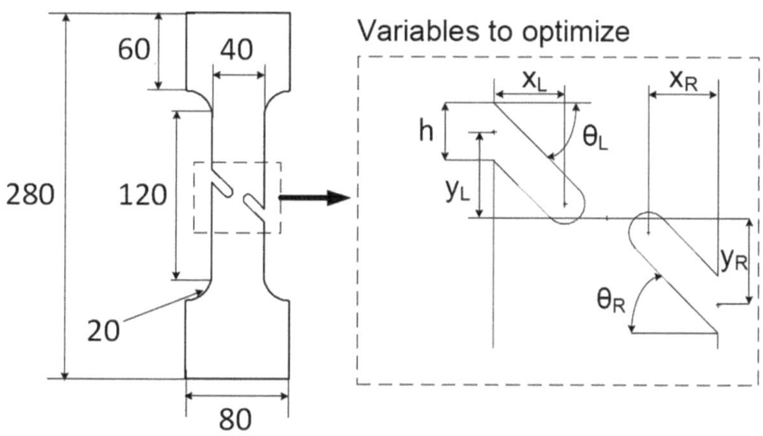

Fig. 12.2 Geometry of the specimen and the seven parametrized independent design variables

Table 12.1 Design variables' values assumed in the 50 configurations studied during the optimization process

Configuration ID	h	θ_L	θ_R	x_L	y_L	x_R	y_R
1	10	−45	45	12.50	15	12.50	−15
2	5	−45	45	12.50	15	12.50	−15
3	10	−30	30	12.50	15	12.50	−15
4	5	−45	45	12.50	20	12.50	−20
5	10	−30	30	12.50	25	12.50	−25
6	5	−30	30	12.50	15	12.50	−15
7	10	−45	45	6	15	6	−15
8	5	−45	45	6	15	6	−15
9	10	−30	30	6	15	6	−15
10	5	−30	30	6	15	6	−15
11	5	−30	30	2.50	5	2.50	−5
12	5	−30	30	2.50	10	2.50	−10
13	10	−30	30	2.50	10	2.50	−10
14	10	0	0	6	15	6	−15
15	5	0	0	6	15	6	−15
16	5	0	0	12.50	10	12.50	−10
17	5	0	0	6	10	6	−10
18	5	0	0	2.50	0	2.50	0
19	5	0	0	2.50	5	2.50	−5
20	5	0	0	2.50	10	2.50	−10
21	10	0	0	2.50	5	2.50	−5
22	10	0	0	2.50	10	2.50	−10
23	10	0	0	12.50	15	12.50	−15
24	5	0	0	12.50	15	12.50	−15
25	20	0	0	5	0	5	0
26	20	0	0	1	0	1	0
27	5	45	45	2.50	0	2.50	0
28	5	45	45	2.50	5	2.50	−5
29	5	45	45	2.50	10	2.50	−10
30	10	45	45	2.50	0	2.50	−0
31	10	45	45	2.50	5	2.50	−5
32	10	45	45	2.50	10	2.50	−10
33	5	45	45	6	0	6	0
34	5	45	45	6	5	6	−5
35	5	45	45	6	10	6	−10
36	10	45	45	6	0	6	0
37	10	45	45	6	5	6	−5
38	10	45	45	6	10	6	−10
39	5	−45	−45	6	0	6	0
40	5	−45	−45	6	5	6	−5
41	5	−45	−45	6	10	6	−10
42	10	−45	−45	6	0	6	0
43	10	−45	−45	6	5	6	−5
44	10	−45	−45	6	10	6	−10
45	5	−45	−45	2.50	0	2.50	0
46	5	−45	−45	2.50	5	2.50	−5
47	5	−45	−45	2.50	10	2.50	−10
48	10	−45	−45	2.50	0	2.50	0
49	10	−45	−45	2.50	5	2.50	−5
50	10	−45	−45	2.50	10	2.50	−10

A Young's Modulus $E = 200$ GPa and Poisson's Ratio $\nu = 0.3$ are input in the FE model for the elastic properties, while the used parameters for the Swift's law are $K = 1000$, $\varepsilon_0 = 0.02$ and $N = 0.5$.

The numerical images are generated to reproduce an experimental set-up equipped with a CCD camera having 1280× 1024 resolution with 8-bit sensor, mounting a lens of 50 mm. The camera is supposed to be mounted with a distance $z = 557$ mm respect to the specimen surface. Moreover, experimental uncertainties are included in the simulated images, assuming that: noise is distributed following an Extreme Value distribution (the Extreme value parameters are function of the GL), the out-of-plane motions are regulated by two angles $\theta = 0.0026$ rad, $\gamma = 0.0026$ rad and the distance $\delta_z = 0.058$ mm, while the in plane motions by $cx = -2.50$ pixel and $cy = -21.40$ pixel. Concerning the lighting conditions, light spot illumination is introduced. See [12] for further details concerning the considered parameters.

12.4 Identification Procedure

In order to retrieve the strain fields required by the inverse identification, the set of synthetic images produced numerically with the simulator is correlated by the MatchID DIC software [20], performing the analysis with a subset size of 23 pixels and a step size of 3 pixels. Hence, the strain field is derived setting a strain window of three points, to reduce filtering to a minimum kernel size equivalent to the subset size.

According with [16] the virtual field the cost function $\Psi(\boldsymbol{\xi})$ is minimized with three different virtual fields:

$$\delta v_1 = \begin{cases} \delta v_x = 0 \\ \delta v_y = y/L \end{cases} \quad \delta v_3 = \begin{cases} \delta v_x = \dfrac{x}{W}\dfrac{(|y| - L)}{L} \\ \delta v_y = 0 \end{cases} \quad \delta v_3 = \begin{cases} \delta v_x = \sin\left(\pi \dfrac{x}{W}\right)\cos\left(\pi \dfrac{y}{H}\right) \\ \delta v_y = \sin\left(\pi \dfrac{x}{W}\right)\cos\left(\pi \dfrac{y}{H}\right) \end{cases} \tag{12.4}$$

where, considering the coordinate system placed with the origin at the centre of the specimen, L is the semi-length of the measurement area and W is the semi-width of the middle section. In particular, the first virtual field is chosen so that the external virtual work is equal to the reaction forces R_j obtained from FE analysis at time increment j:

$$\frac{1}{2}\int_{\delta V}\left(\mathbf{T}_j^{1PK}\mathbf{n}_0\right) \cdot \delta v_1^* dS = R_j. \tag{12.5}$$

Differently, the virtual work of the external force, for the second and the third virtual fields of Eq. (12.2), is equal to zero.

12.5 Results and Discussion

The analysis was executed using first directly the strain field resulting from FEA, then three different sets of synthetic images: the former is composed by pure images, the second one involves noise in picture generation, the third set has all the other experimental uncertainties.

The overall results are displayed in Fig. 12.3 in terms of RMSE ($\bar{\varepsilon}_{\text{lim}} = 0.6$) and in terms of material stress-strain curve. The analysis performed on FEM data shows results close to the ideal material behavior, reaching, for the best geometry (configuration 39), a RMSE of 0.21 %. Figure 12.4a depicts the PEEQ distribution for that configuration. The geometries with higher error are characterized by a reduced strain distribution.

When the DIC process is simulated, the RMSE increases as expected. Using pure images, the resulting best geometry is the same as the one obtained from FEM data. Concerning the second set of numerical images, adding noise deteriorates the pictures and, consequently, the identification. Nonetheless, the best geometry is still in agreement with the previous results, although its RMSE is about 20 MPa. Including all the experimental error sources strongly affects the identification procedure. In fact, the best geometry changes respect to the previous cases. The RMSE value for previous best geometry (configuration 39 in Table 12.1) almost exceeds 200 MPa, while shows the lowest value for configuration 14, which relative PEEQ map is illustrated in Fig. 12.4b.

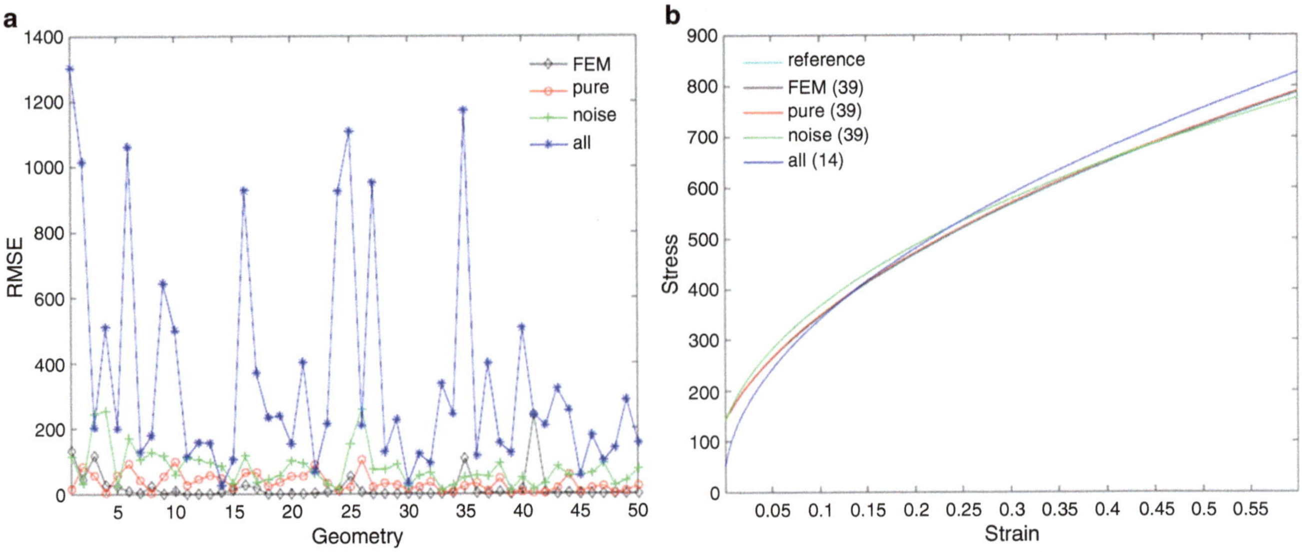

Fig. 12.3 (**a**) Identification error for the 50 configurations respect to FEM and the 3 sets of numerical images. (**b**) Comparison of different stress-strain curves from geometries having the best identification performances

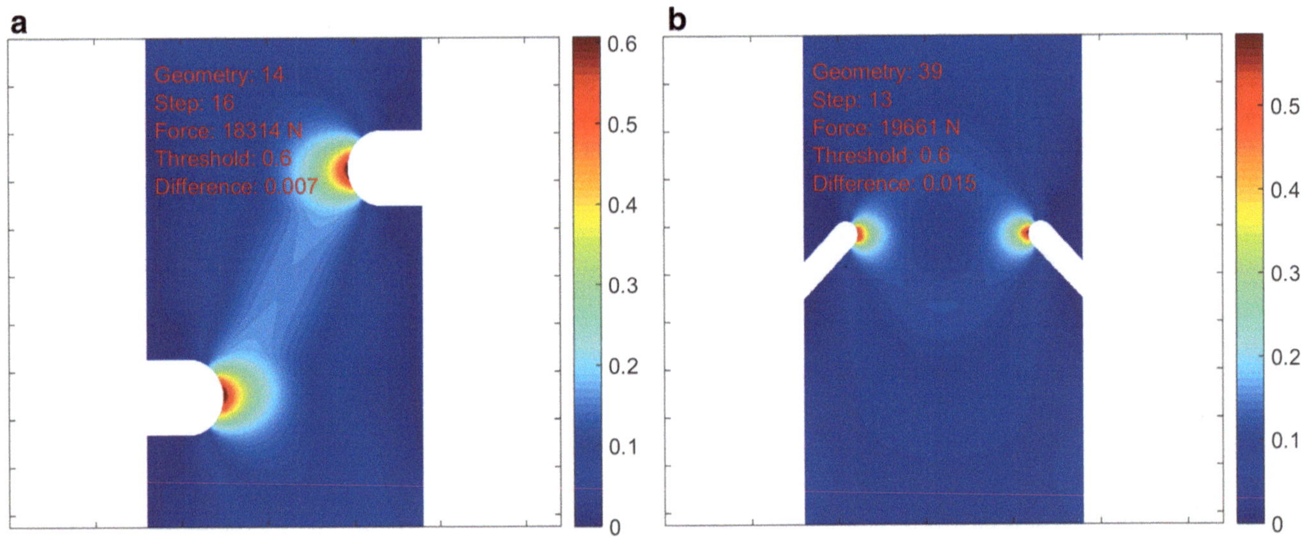

Fig. 12.4 Comparison of the optimal geometry obtained from identification on FEM data (**a**) and identification on synthetic images with all experimental error sources (**b**), respectively. The maps display the equivalent plastic strain at the final time increment

12.6 Conclusion

The present paper introduces a numerical procedure to optimize the shape of specimens involved in the identification of hardening behaviour of metals using the non-linear VFM. Starting from 50 geometry configurations described by 7 independent variables and employing a numerical simulator that generates synthetic images, the optimal geometry is assessed looking at the identified stress-strain curve. Different simulated experiments were performed considering FEA data, simulated DIC measurement and simulated DIC measurement with simulated experimental errors. The analysis of results shows that the optimization method is capable to distinguish adequate configurations from not-optimal ones. As general remarks, specimens with better performance are the ones characterized by a diffused plastic strain area, while localized strain concentrations in proximity of notches produce worse identification.

In future, this approach will be extended to other specimen's shape and to more complex constitutive models, or improved using more advanced optimization methods, for instance, as genetic algorithm.

References

1. Grédiac, M., Hild, F., Pineau, A.: Full-Field Measurements and Identification in Solid Mechanics. Wiley, New York (2012)
2. Avril, S., Bonnet, M., Bretelle, A.-S., Grédiac, M., Hild, F., Ienny, P., Latourte, F., Lemosse, D., Pagano, S., Pagnacco, E., Pierron, F.: Overview of identification methods of mechanical parameters based on full-field measurements. Exp. Mech. **48**, 381–402 (2008)
3. Sutton, M.A., Orteu, J.J., Schreier, H.W.: Image Correlation for Shape, Motion and Deformation Measurements. Springer, New York (2009)
4. Pierron, F., Grédiac, M.: The Virtual Fields Method. Springer, New York (2012)
5. Lava, P., Cooreman, S., Coppieters, S., De Strycker, M., Debruyne, D.: Assessment of measuring errors in DIC using deformation fields generated by plastic FEA. Opt. Laser Eng. **47**, 747–753 (2009)
6. Lava, P., Cooreman, S., Coppieters, S., De Strycker, M., Debruyne, D.: Study of systematic errors in strain fields obtained via DIC using heterogeneous deformation generated by plastic FEA. Opt. Laser Eng. **48**, 457–468 (2010)
7. Wang, Y.Q., Sutton, M.A., Brook, H.A., Schreier, H.W.: Quantitative error assessment in pattern matching: effects of intensity pattern noise, interpolation, strain and image contrast on motion measurements. Strain **45**, 160–178 (2009)
8. Rossi, M., Pierron, F.: On the use of simulated experiments in designing tests for material characterization from full-field measurements. Int. J. Solids Struct. **49**, 420–435 (2012)
9. Avril, S., Vautrin, A., Surrel, Y.: Grid method: application to the characterization of cracks. Exp. Mech. **44**, 37–43 (2004)
10. Rossi, M., Lava, P., Pierron, F., Debruyne, D., Sasso, M.: Effect of DIC spatial resolution, noise and interpolation error on identification results with the VFM. Strain **51**, 206–222 (2015)
11. Wang, P., Pierron, F., Rossi, M., Lava, P., Thomsen, O.: Optimized experimental characterization of polymeric foam material using DIC and the virtual fields method. Strain **52**(1), 59–79 (2016)
12. Badaloni, M., Rossi, M., Chiappini, G., Lava, P., Debruyne, D.: Impact of experimental uncertainties on the identification of mechanical material properties using DIC. Exp. Mech. **55**, 1411–1426 (2015)
13. Grédiac, M., Pierron, F.: Applying the virtual fields method to the identification of elasto-plastic constitutive parameters. Int. J. Plast. **22**, 602–627 (2006)
14. Rossi, M., Pierron, F.: Identification of plastic constitutive parameters at large deformations from three dimensional displacement fields. Comput. Mech. **49**, 53–71 (2012)
15. Kim, J.H., Barlat, F., Pierron, F.: Determination of anisotropic plastic constitutive parameters using the virtual fields method. Exp. Mech. **54**, 1189–1204 (2014)
16. Rossi, M., Barlat, F., Pierron, F., Sasso, M., Lattanzi, A.: Identification of the YLD2000-2D model with the virtual fields method. In: Conference Proceedings of the Society for Experimental Mechanics Series, vol. 9, pp. 51–57 (2016)
17. Kim, J.K., Serpantié, A., Barlat, F., Pierron, F., Lee, M.G.: Characterization of the post-necking strain hardening behavior using the virtual fields method. Int. J. Solid Struct. **50**, 3829–3842 (2013)
18. Broggiato, G., Campana, F., Cortese, L., Mancini, E.: Comparison between two experimental procedures for cyclic plastic characterization of high strength steel sheets. Eng. Mater. Technol. **134**, 63–72 (2012)
19. Cortese, L., Coppola, T., Campanelli, F., Broggiato, G.: A J2-J3 approach in plastic and damage description of ductile materials. Int. J. Damage Mech. (2015)
20. http://www.matchidmbc.be/

Chapter 13
New Methodology for Steady-State Friction Measurements of Granular Materials Under Pressure

M. Bek, J. Gonzalez-Gutierrez, and I. Emri

Abstract Flow of granular material during processing, handling and transportation strongly influences the quality of the final product and its cost, that is why it is important to measure flow properties of granular materials. Flowability of granular materials depends on the characteristics of the material and on the conditions at which flow is occurring. In this paper a new methodology is introduced to measure friction between granular materials under pressure induced with uniaxial compression. Apparatus also allows analysis of conditions at which granular material starts to flow when exposed to uniaxial compressive load, i.e., zero-rate flowability. We call the apparatus the Granular Friction Analyzer (GFA).

The concept of the GFA was tested by measuring four different materials with different average particle sizes. It was observed that as the particle size decreases so does its zero-rate flowability. This is in agreement with powder literature. Therefore, it can be concluded that in general the GFA method can be a very useful tool to study friction between granular materials and conditions at which the granular material flow initiates, i.e. zero-rate flowability of powders under pressure.

Keywords Granular materials • Inherent friction • Zero-rate flowability • Granular flow • Metal powders • Uniaxial compression

13.1 Introduction

Granular materials are present in our everyday life whether as raw materials, intermediate or final products. For instance, one finds them in the food industry from cereal grains to dehydrated milk; in the pharmaceutical industry different powders are blended and compressed to produce tablets; metal, polymeric or ceramic parts can be produced through different powder sintering techniques; moreover, granular materials are present in nature as sand and rocks. In all these examples, the flow of powder during processing, handling and transportation strongly influences the quality of the product and its cost.

A granular material is defined as a collection of discrete solid particles that are filled with an interstitial fluid, usually air. From this definition, one can notice that they cannot be classified either as solids or liquids, leading to a complex behavior difficult to understand and predict [1]. In the literature, several terms are used interchangeably to describe granular materials, such as bulk solids, particulate solids and powder.

Within the world of granular materials, metal powders constitute an important part of the industry. In this context, one important topic of research nowadays is to study the flowability of powders in powder metallurgy because an increase in flowability represents an increase in the productivity of the overall process and a reduction in the cost of transportation of powders. During powder metallurgy granular materials are transported and molded under pressure.

Another example of utilizing granulated materials is in new emerging dissipative granular technology [2, 3]. In this technology damping elements are made of granulated polymeric materials and encapsulated in a flexible container. Good performance of such damping elements can be achieved by pressurizing granular material to a pressure level, at which the damping element's frequency for absorbing a maximum energy substantially matches the frequency of mechanical excitation. This is possible because damping properties of polymeric (elastomeric) materials are frequency (time), temperature and pressure dependent [4, 5]. In this instance also flowability of polymeric particles is important, since increased flowability ensures the equal distribution of pressure inside damping elements.

M. Bek (✉) • I. Emri
Faculty of Mechanical Engineering, University of Ljubljana, Pot za Brdom 104, Ljubljana, Slovenia
e-mail: bek.marko@gmail.com

J. Gonzalez-Gutierrez
Faculty of Mechanical Engineering, University of Ljubljana, Pot za Brdom 104, Ljubljana, Slovenia

Montanuniversität Leoben, Otto Glöckel-Straße 2, Leoben, Austria

Y. Zhu, A.T. Zehnder (eds.), *Experimental and Applied Mechanics, Volume 4*,
Conference Proceedings of the Society for Experimental Mechanics Series, DOI 10.1007/978-3-319-42028-8_13

Flowability is defined as the ability of a powder to flow and results from the combination of material physical properties, which along with the equipment used for handling or processing governs how easy a granular material can flow. There are several techniques used for measuring flowability some of them require simple apparatuses while others use sophisticated setups. Simple traditional apparatuses as Hall-flowmeters and angle of repose devices provide a comparative analysis of which particular powder flows better with respect to another under certain conditions of measurement [6, 7]. On the other hand, more sophisticated setups as shear cells and powder rheometers can provide more quantitative information as cohesion and angle of friction that influence the flowability of powders [8, 9]. Nevertheless since flowability changes according to the conditions at which flow is occurring, measuring devices more or less should resemble the handling or processing equipment in which the powder will be used [10]. An overview of the flow properties of the granular systems including voids, granular porosity and random packing characteristics is summarized in the paper of Traina et al. [11]. Since most of the measuring techniques available in the market work under gravitational and shear flows, the present paper aims to introduce a new method of characterization where the driving force for granular flow is high pressure, which was recently published [12]. Hence, apparatus allows investigation of the so-called zero-rate flowability induced by pressure. More precisely the apparatus allows investigation of friction within granular materials exposed to pressure generated by uniaxial compression. As a special case apparatus allows determination of conditions at which granular material will start to move.

13.2 Materials and Methods

Different equipment available in the market for measuring flowability of granular material are based on gravitational forces or shearing failure. The Granular Friction Analyzer (GFA) is a novel apparatus for studying inherent friction in granular materials that are exposed to uniaxial compression loading and conditions at which flow of granular material will be initiated—zero-rate flowability. The apparatus was developed at the Center for Experimental Mechanics (CEM), Faculty of Mechanical Engineering, University of Ljubljana.

Figure 13.1 shows schematically the main parts of the apparatus. The GFA is composed of a hollow cylinder with one of its ends closed that is loaded with granular material. Once the cylinder is filled with particles, a force is applied by means of a piston (Fig. 13.1a); such force will induce a pressure inside the cylinder that at the same time induces elastic deformations in the axial and tangential directions of the external wall of the cylinder. The strain in the external face of the cylinder can be measured experimentally, e.g., by means of a strain gages setup (Fig. 13.1b), or optically. Finally, Fig. 13.1c shows the dimensions of the cylinder. One can notice that the powder inside the cylinder is subjected to a confined compression;

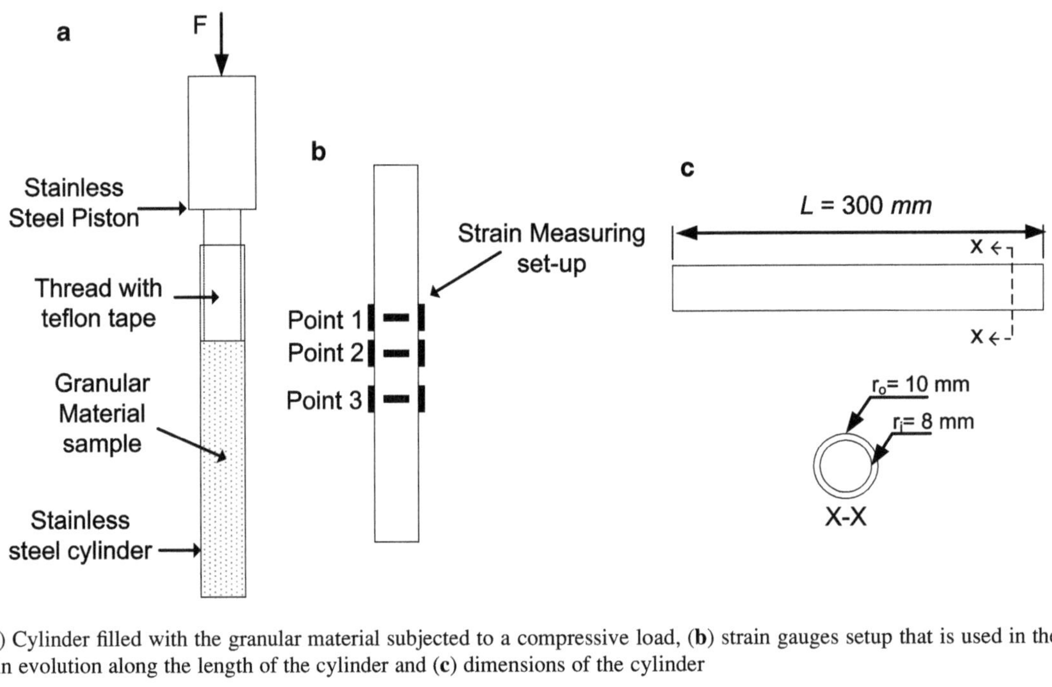

Fig. 13.1 (**a**) Cylinder filled with the granular material subjected to a compressive load, (**b**) strain gauges setup that is used in the apparatus to measure strain evolution along the length of the cylinder and (**c**) dimensions of the cylinder

Ma and Ravi-Chandar [13] developed an apparatus that works under the same principle, however such compression was applied to solid materials and their goal was to characterize the stress state of the inserted solid material.

The goal of the GFA is to obtain the pressure profile along the length of the cylinder and to introduce such profile as a measure of inherent friction between particles of granular materials. The concept of the GFA is valid because forces through the granular material and the pressure distribution along the length of the cylinder is a measure of how force is transmitted from the piston to the powder and from the powder to the cylinder. In this context, one expects that depending on the friction properties of the powder, the pressure will be transmitted in a different way along the cylinder, thus allowing us to obtain a method for determining inherent friction between particles and condition at which the granular material will start to move. By applying force at different rates the GFA may be used for studying inherent friction between granular materials in dynamic-equilibrium as well. Such information is of importance for pharmaceutical industry in production of tablets, for example.

13.2.1 Working Principle of GFA

Pressure distribution in axial direction is the final result that we seek from the apparatus, however what it is being measured experimentally are strains in the tangential (ε_θ) and axial (ε_z) directions along the length of the cylinder. Figure 13.2 show schematically the expected results for the strain evolution in a Newtonian fluid, and in a granular material, respectively. In the case of the fluid (Fig. 13.2a), it is known that the hydrostatic pressure is constant along the length of the cylinder and it is proportional to the tangential strain, whereas the axial strain occurs just due to Poison effect. In the case of the granular material (Fig. 13.2b), it is expected that the pressure is maximal close to the piston where the force is applied, followed by a pressure drop. The pressure drop occurs because the force is distributed through layers of particles and there is formation of force chains that attenuate the pressure along the cylinder and increase the friction force between particles and inner wall of the cylinder [14, 15]. In this case, the tangential strain is maximal at the position where the force is applied and drops in the same way as pressure does, whilst the axial strain increases as the tangential decreases because the force is been transmitted to the wall as an axial load due to friction. In powder technology the relation between axially applied pressure and pressure in radial direction is known as 'Janssen effect' [16]. For more details, we refer the reader to the paper of Ovarlez et al. [17].

Equation to obtain the pressure along the axial direction of the cylinder as a function of the measured strains in tangential and axial direction is expressed as:

$$p_i = \frac{\left(\varepsilon_{\theta(m)} - \nu\varepsilon_{z(m)}\right)E\left(r_0^2 - r_i^2\right)}{2r_i^2} \tag{13.1}$$

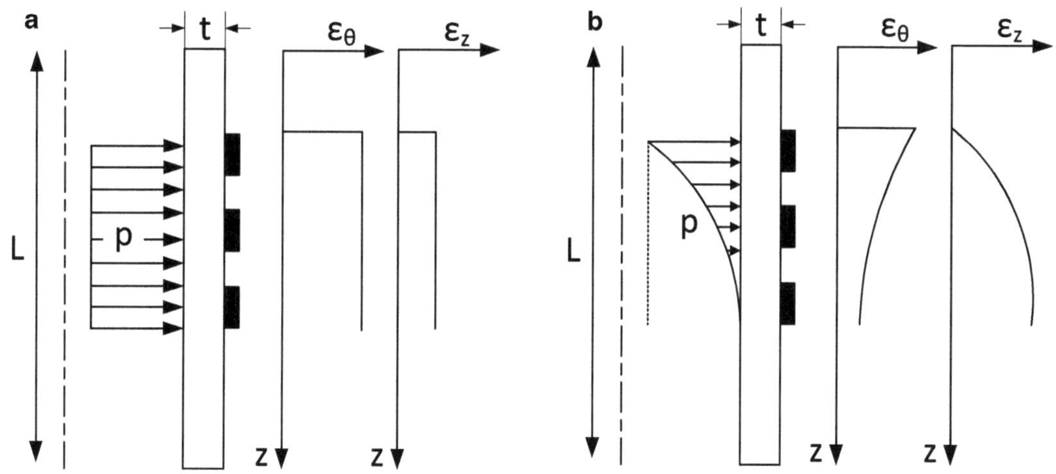

Fig. 13.2 Expected results for tangential and axial strain along the length of the cylinder of the GFA (**a**) for a fluid and (**b**) for a granular material

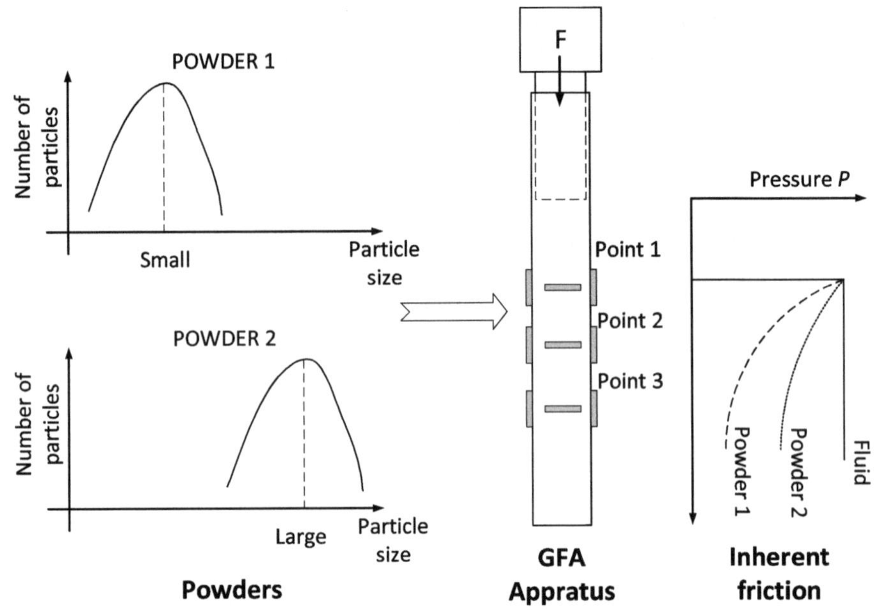

Fig. 13.3 Measuring capabilities of the GFA apparatus

Figure 13.3 illustrates how one can compare the inherent friction properties two granular materials in the GFA, taking as reference the pressure profile of a fluid which is constant along the length of the cylinder, one can determine if certain powder has higher or lower inherent friction by analyzing how steep is the pressure drop along the length of the cylinder, i.e. a granular material with less pressure drop will lower inherent friction (behavior closer to liquid) than another with very rapid pressure drop. Based on the literature, it is expected that powders with a smaller average particle size will exhibit higher inherent friction [14, 15]; for this reason Powder 1 in Fig. 13.3 has a much faster pressure drop than Powder 2.

13.2.2 GFA Engineering Solution

The GFA can be subdivided into two systems: (i) Mechanical system, and (ii) strain measuring system. The mechanical system of the apparatus is composed by the cylinder, piston, measuring plates and a fixation frame that supports and couples the previous components. In this case the strain measuring system is composed of strain gages that are glued on the external face of the cylinder and the completion circuit (one dummy and two resistors placed on a "measuring plate" that is screwed to the fixation frame) that are required to complete the "quarter Wheatstone bridge with temperature compensation" that is used in all circuits. For higher precision, strain gages may be replaced with a more versatile optical strain measuring system. To complete the strain measuring system, the Wheatstone bridges must be coupled to a signal conditioning system that powers the circuits and allows us to record the measuring signal.

Figure 13.4 shows schematically the complete setup of the GFA measuring process in a flow chart form. First a force is transmitted to the granular material by means of the piston, and this force induces an internal pressure that leads to strain in the outer face of the cylinder. When the strain gages in the cylinder are connected to a Wheatstone bridge a signal of voltage that is related to a strain in the cylinder is recorded. The force sensor and the Wheatstone bridges in the strain measuring system are connected to the amplifier/signal conditioning that powers the circuits and increases the output signal to a recordable level and minimizes the noise in the signal; finally the analog signal from the amplifier is converted into a digital format through a data acquisition (DAQ) card and then is stored in a PC through a LabVIEW (National Instruments, Austin, TX, USA) interface.

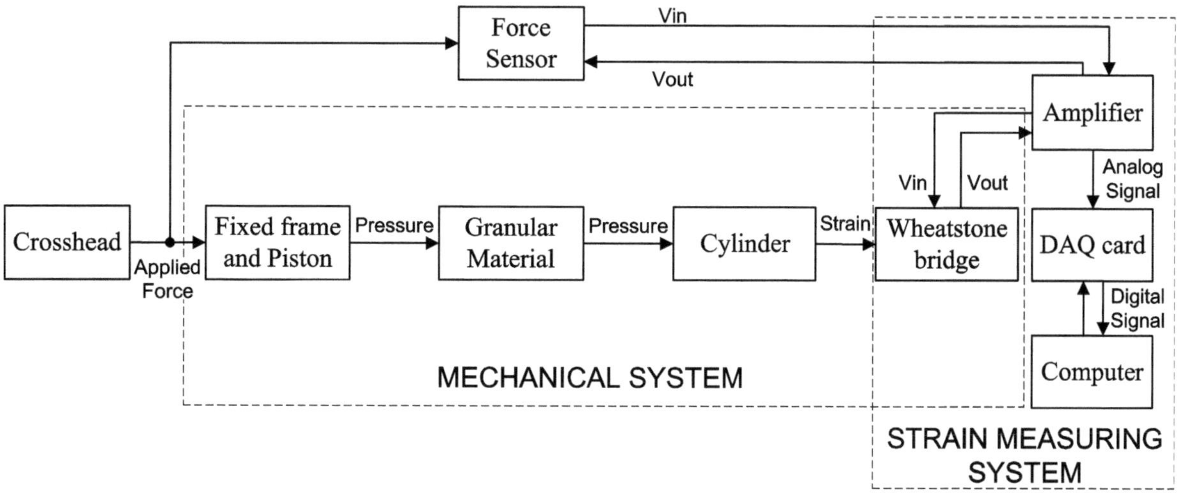

Fig. 13.4 Mechanical and strain measuring system of the GFA, flowchart of the measuring process

Table 13.1 Average particle size of materials tested

Material identification	Type of material	Average particle size [μm]
Fraction I	Stainless steel 316LW spherical powder	41.7
Fraction II		23.7
Fraction III		20
Fraction IV		6.8
POM	Polyoxymethylene copolymer spherical pellets	4000

13.3 GFA Apparatus Performance Demonstration

For the purpose of GFA apparatus performance demonstration five different materials provided by BASF (Germany) were analyzed: four stainless steel powders and one type of polyoxymethylene (POM) pellets. In Table 13.1, their average particle size is shown. In this demonstration we limit ourselves to study the effect of the average particle size only. Behavior of granular material is of course affected by grains size distribution, grains shape, grains surface roughness and cohesive forces which however are not the subject of the current paper.

13.3.1 Methodology of Experiments

The measuring procedure is divided in four stages. First, the powder sample should be prepared and consolidated by tapping the cylinder. It is important to point out the importance of sample conditioning as part of the sample preparation process in order to prevent excess of air in between the particles. To release the excess of air from the sample and yield the same arrangement of particles inside the cylinder, tapping of the cylinder is necessary during and after filling the cylinder with powder: this operation must be done as repeatable as possible. The second stage consists of placing the filled cylinder into the frame of the apparatus, and balancing of the circuits. During the third stage, the test is performed by applying the load. The final stage is the cleaning of the cylinder after the test.

Before performing any test, one has to determine the right amount of powder sample to be used during the experiments, this is done by performing preliminary experiments that follow the same steps described, but with the only difference that the amount of powder is changed until satisfying the condition that the position of the piston after applying the desired load is about 17 mm above the first measuring point (in our case 103.5 mm from the top of the cylinder). In order to determine the height to which the powder should be loaded, Finite Element Analysis (FEA) was performed. The analysis was performed by applying 30 MPa of hydrostatic pressure along the distance of 225 mm, measured from the bottom of a 300 mm stainless steel cylinder. From FEA, it was observed that due to the fixed end at the bottom of the cylinder, there is an area of tangential

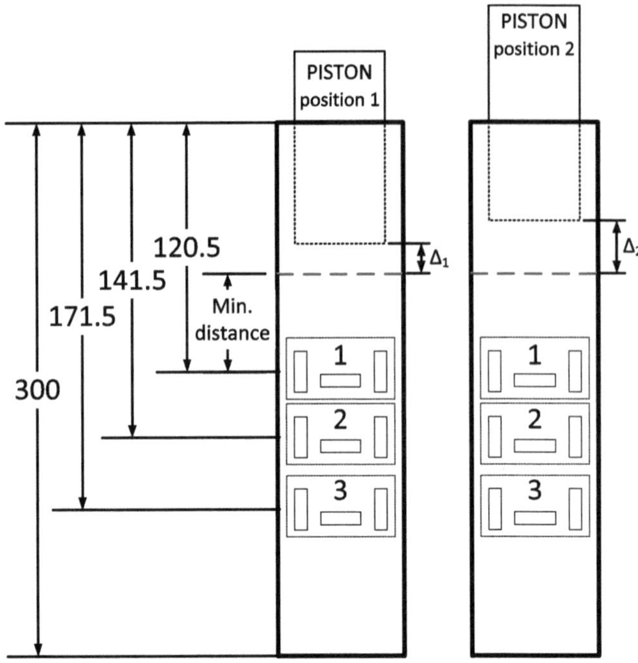

Fig. 13.5 Position of piston for determination of pressure profiles

strain instability at about 20 mm from the bottom of the cylinder and at about 15 mm below the point where the hydrostatic pressure is being applied. Therefore, the distance of 17 mm from the first measuring point was chosen in order to avoid the instability zone.

All materials were tested ten times following the described protocol. In order to artificially increase the number of measuring points along the length of the cylinder, the amount of powder sample was deliberately changed and in this particular case two different distances were used as schematically shown in Fig. 13.5. When determining the measuring position the minimal distance of 15 mm from the first measuring point has to be taken into account in order to avoid the instability zone. Using this procedure pressure profile curves composed of six points instead of just three were obtained. The force that was applied during the tests was $F = 4428 \pm 25$ N corresponding to a maximum pressure $p = 22.025 \pm 0.125$ MPa.

13.3.2 The GFA Index

The inherent friction of the granular material is represented by axial distribution of pressure within the measuring cylinder. If the cylinder is filled with a Newtonian fluid the pressure would be uniform throughout the volume of the measuring cylinder, i.e., constant along the axial direction, as schematically shown in Fig. 13.3. On the other hand, when the cylinder is filled with a granular material pressure distribution is not constant, it is changing along the axial direction of the cylinder, as shown in Fig. 13.3. For easy comparison pressure distribution along the axial direction of the cylinder may be normalized by the pressure, which would be generated within the cylinder if it would be filled with a Newtonian fluid, hence $p_{NF} = F/A$. Such two situations are shown more vividly in Fig. 13.6. As a measure of the inherent friction we may define a parameter, which we will call "GFA" index, and it is defined as

$$GFA_{index} = \frac{S_{granular}}{S_{fluid}} = \frac{\int_0^L p(z)dz}{p_{NF} \cdot L} = \frac{A}{FL} \cdot \int_0^L p(z)dz, \tag{13.2}$$

where $S_{granular}$ is the area of the pressure profile of a granular material, and S_{fluid} is the area of the pressure profile corresponding to a Newtonian fluid, F is the applied force, A is the inner cross section area of the cylinder, $p(z)$ is the

Fig. 13.6 Schematic illustration and definition of the "GFA index"

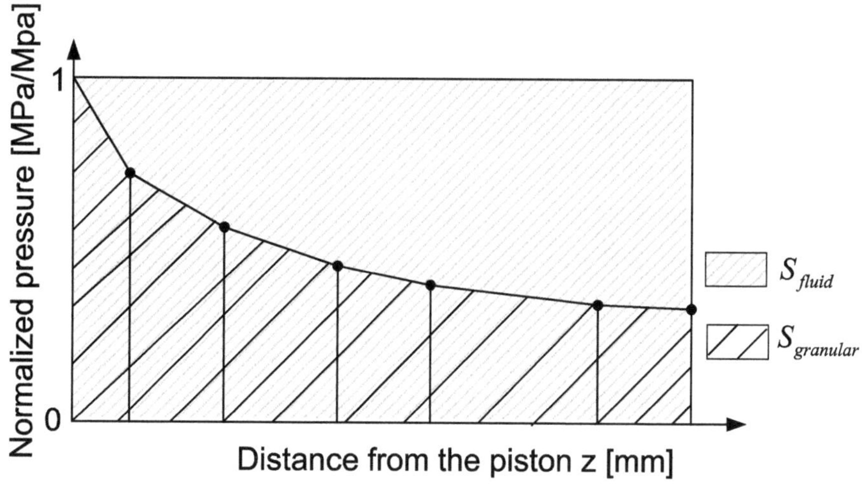

pressure distribution along the cylinder and L is an agreed upon "standardized" length of the cylinder. In our case we have used $L = 188$ mm.

The maximum value of the "GFA index" is 1 and indicates that tested granular material does not exhibit any inherent friction as a Newtonian fluid. As the "GFA index" decreases it indicates amplification of inherent friction in a granular material. In this case the area may be calculated by using trapezoidal rule, as it is shown in Fig. 13.6 where the total area of the curve corresponds to the sum of six trapezoids that are formed from the number of measuring points used to calculate the pressure profile. By increasing the number of measured points determination of the GFA index will be of course improved.

13.4 Results and Discussion

Following the methodology previously described, the pressure profiles of all metallic powders and POM material were measured. In line with the definition of GFA index measured pressures were normalized by the maximal pressure, which corresponds to the applied force divided by the cross section area of the cylinder. The maximum pressure varied slightly from test to test within the range from 21.9 to 22.2 MPa.

Figure 13.7a shows the "GFA index" for five granular materials as function of the average particle size. In Fig. 13.7a we have used logarithmic scale for the particle size in order to show all five results on the same diagram. Figure 13.7b shows the "GFA index" for four metallic granular materials only. Here the particles size coordinate is shown in linear scale. The results shown represent average of ten measurements, and the corresponding variance was calculated by using Student's t-distribution.

From the results one can observe that Fraction IV presents the lowest "GFA index" which indicates that it has the highest inherent friction. This agrees with [14, 15, 18] where authors state that as particle size decreases cohesion increases and thus inherent friction increases. Fractions III and II present almost the same GFA index and Fraction I has the highest "GFA index" indicating the lowest inherent friction among the metal powders. Despite the large experimental error observed in measurement on POM, the results evidence that POM have much lower inherent friction with respect to metal powders because of the lower friction between the POM pellets. Large experimental error in POM measurements may be accounted to unfavorable ratio between the size of the POM pellets and the inner diameter of the testing cylinder.

We can observe that the value of the "GFA index" of POM is approximately three times higher than the value of the "GFA index" for the metallic powders, however, it is important to notice that the difference in average particle size is more than 100 times, see Table 13.1.

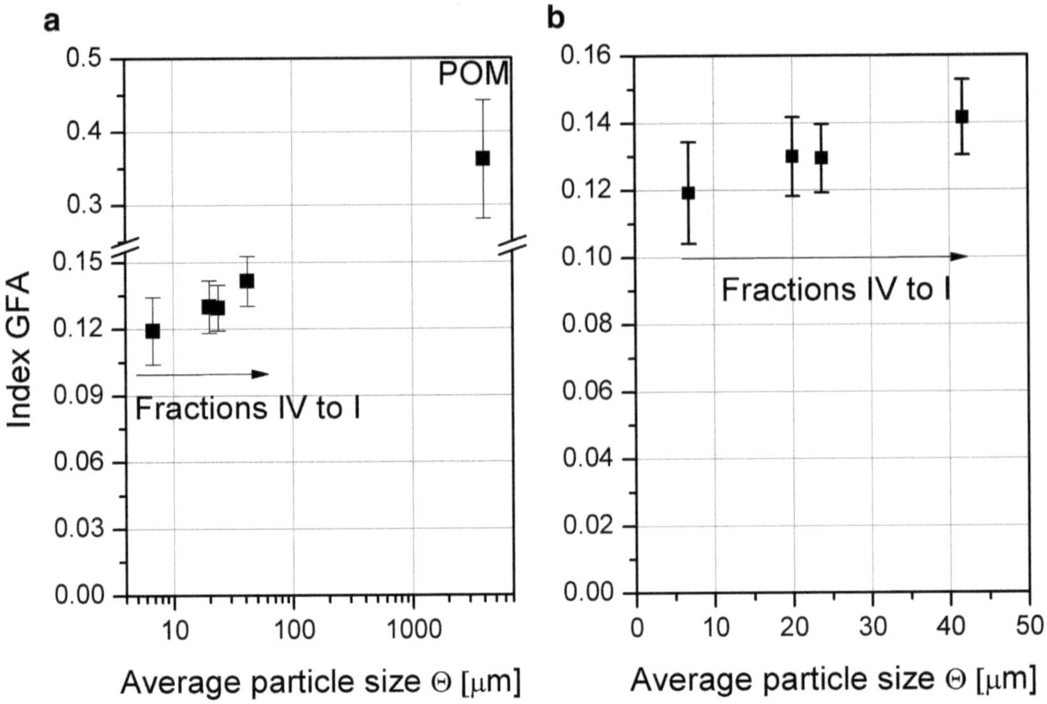

Fig. 13.7 (**a**) GFA index for metal powders and POM, (**b**) GFA index for metal powders (close up look)

13.5 Conclusions

Proposed GFA apparatus proves to be an interesting tool for studying the inherent friction of granular materials exposed to pressure loading and conditions at which granular material will start to flow. It consists of a hollow cylinder where granular materials can be pressurized by means of a piston; the pressure drop in axial direction of the cylinder is estimated by measuring strains at the outer surface of the cylinder along its length. A new parameter, GFA-index, used to assess the inherent friction of different granular materials has also been introduced. The GFA-index is the ratio of the integral of the pressure distribution along the agreed upon "standardized" length of the cylinder, L, divided by the integral over the same distance assuming that pressure is constant, which is the case if cylinder would be filled with a Newtonian fluid.

The GFA prototype was utilized to test five fractions of stainless steel 316LW and granules of polyoxymethylene of different average particle size in order to demonstrate the concept and capabilities of the apparatus. Obtained results proved the simplicity and power of the GFA apparatus. Obtained results were in line with the data from the literature. However, demonstration experiments have shown that measuring strains at only few locations is not sufficient, which suggests that further work is required to improve the accuracy of this first GFA prototype.

Important recommendations towards improving the accuracy of the GFA include: standardization of conditions during the sample preparation (such as powder filling and tapping of the cylinder); the implementation of a sensor to control the displacement of the piston during the test; and measuring strains at the outer surface with an optical method that allows obtaining the complete pressure profile along the cylinder instead at just three positions. The most appealing alternative is to use speckle interferometry.

Acknowledgements Authors acknowledge the financial support of the Slovenian Research Founding Agency, and the European Union Social Fund. As well as the in kind support from BASF SE, Germany.

References

1. Jaeger, H.M., Nagel, S.R.: Physics of the granular state. Science **255,** 1523–1531 (1992). doi:10.1126/science.255.5051.1523
2. Emri, I., von Bernstorff, B.S., Brehmer, F., Kalamar, A., Bek, M., Oblak, P.: Sleeper with damping element based on dissipative bulk or granular technology. EP2700838 (2012)

3. Emri, I., von Bernstorff, B.S.: Dissipative bulk and granular systems technology. EP2700839 (2012)
4. Tschoegl, N., Knauss, W., Emri, I.: The effect of temperature and pressure on the mechanical properties of thermo-and/or piezorheologically simple polymeric materials in thermodynamic equilibrium—a critical review. Mech. Time-Depend. Mater. **6**, 53–99 (2002). doi:10.1023/A:1014421519100
5. Emri, I., Gonzalez-Gutierrez, J., Gergesova, M., Zupančič, B., Saprunov, I.: Experimental determination of material time-dependent properties. In: Hetnarski, R. (ed.) Encycl. Therm. Stress, pp. 1494–1510. Springer, Netherlands (2014). doi:10.1007/978-94-007-2739-7_907
6. Vlachos, N., Chang, I.T.H.: Investigation of flow properties of metal powders from narrow particle size distribution to polydisperse mixtures through an improved Hall-flowmeter. Powder Technol. **205**, 71–80 (2011). doi:10.1016/j.powtec.2010.08.067
7. Geldart, D., Abdullah, E.C., Hassanpour, A., Nwoke, L.C., Wouters, I.: Characterization of powder flowability using measurement of angle of repose. China Particuology **4**, 104–107 (2006). doi:10.1016/S1672-2515(07)60247-4
8. Shinohara, K., Golman, B.: Dynamic shear properties of particle mixture by rotational shear test. Powder Technol. **122**, 255–258 (2002). doi:10.1016/S0032-5910(01)00422-3
9. Freeman, R.: Measuring the flow properties of consolidated, conditioned and aerated powders—a comparative study using a powder rheometer and a rotational shear cell. Powder Technol. **174**, 25–33 (2007). doi:10.1016/j.powtec.2006.10.016
10. Prescott, J., Barnum, R.: On powder flowability. Pharm. Technol. **24**, 60–84 (2000). citeulike-article-id:11996698. http://cat.inist.fr/?aModele=afficheN&cpsidt=795318
11. Traina, K., Cloots, R., Bontempi, S., Lumay, G., Vandewalle, N., Boschini, F.: Flow abilities of powders and granular materials evidenced from dynamical tap density measurement. Powder Technol. **235**, 842–852 (2013). doi:10.1016/j.powtec.2012.11.039
12. Bek, M., Gonzalez-Gutierrez, J., Moreno Lopez, J.A., Bregant, D., Emri, I.: Apparatus for measuring friction inside granular materials—granular friction analyzer. Powder Technol. **288**, 255–265 (2016). doi:10.1016/j.powtec.2015.11.014
13. Ma, Z., Ravi-Chandar, K.: Confined compression: a stable homogeneous deformation for constitutive characterization. Exp. Mech. **40**, 38–45 (2000). doi:10.1007/BF02327546
14. Campbell, C.S.: Granular material flows—an overview. Powder Technol. **162**, 208–229 (2006). doi:10.1016/j.powtec.2005.12.008
15. Rao, K.K., Nott, P.R.: An Introduction to Granular Flow. Cambridge University Press, New York (2008). doi:http://dx.doi.org/10.1017/CBO9780511611513
16. Janssen, H.A.: Versuche über getreidedruck in silozellen. Zeitschr. D. Vereines Dtsch. Ingenieure. **39**, 1045–1049 (1895)
17. Ovarlez, G., Fond, C., Clément, E.: Overshoot effect in the Janssen granular column: a crucial test for granular mechanics. Phys. Rev. E **67**, 60302 (2003). doi:http://dx.doi.org/10.1103/PhysRevE.67.060302.
18. Schulze, D.: Powders and Bulk Solids: Behavior, Characterization, Storage and Flow. Springer, New York (2007)

Chapter 14
Constitutive Response of AA7075-T6 Aluminum Alloy Sheet in Tensile and Shear Loading

T. Rahmaan, C. Butcher, and M.J. Worswick

Abstract Tensile and shear experiments were performed on AA7075-T6 sheet at strain rates ranging from quasi-static ($0.001\ s^{-1}$) to high ($1000\ s^{-1}$) in three sheet orientations (0°, 45° and 90°) with respect to the rolling direction. Digital image correlation (DIC) techniques were employed to measure the strains in the experiments. The AA7075-T6 alloy showed mild rate sensitivity over the range of strain rates tested. The level of plastic anisotropy was characterized and was shown to be rate-insensitive. The quasi-static experimental data was used to calibrate the eight-parameter Barlat YLD2000 anisotropic yield criterion to describe the anisotropic behaviour of the sheet material. The quasi-static hardening behaviour to large strains was also experimentally determined by converting the shear stress to an equivalent uniaxial stress and fit using a Hockett–Sherby model. The calibrated anisotropic yield criterion was able to capture the material anisotropy providing good agreement with the experiment data.

Keywords AA7075-T6 aluminum alloy sheet • Tensile and shear constitutive behaviour • Anisotropy • Strain rate sensitivity • Constitutive fitting

14.1 Introduction

Aluminum alloys are widely used in automotive structural components due to their low density and high strength. In automobile crash events or high speed metal forming, the local strain rates can vary from quasi-static conditions to over $1000\ s^{-1}$ [1], and understanding the constitutive response associated with dynamic loading is critical for the accurate design of sheet metal forming operations of aluminum structural components and to predict their crash performance. In addition to the material hardening behavior, the plastic anisotropy (r-value) of automotive sheet metals also plays an important role determining the magnitude of local deformation. Uniaxial tensile experiments are typically used to characterize the mechanical properties of sheet metals until the onset of necking. However, due to relative limited diffuse necking followed by localization, tensile experiments are limited to relatively small strains, particularly for high strength aluminum alloys such as the 7000-series [2]. As an alternative to tensile tests, shear experiments can be used to investigate the material response under large plastic strains since localized necking does not occur and plane stress conditions are operative until fracture. Aluminum alloys are typically considered to have low strain rate sensitivity behavior at room temperature [3]. 5000-series aluminum alloys exhibit negative rate sensitivity at low strain rates [4] due to PLC band propagation; however, PLC band propagation in AA5182 is suppress at strain rates above $1\ s^{-1}$ at which point positive rate sensitivity is established [5] with increases in tensile elongation [4]. Studies of the effect of strain rate and different loading conditions on the constitutive behavior of AA7075-T6 alloy sheet are limited; however, Nicholas [6] conducted tensile experiments at strain rates up to $10^3\ s^{-1}$ on AA7075, and found low but measurable strain rate sensitivity in tension. On the other hand, El-Magd and Abouridouane [7] found that the flow stress increases slightly with strain rate.

In the current work, the low (0.001–$0.01\ s^{-1}$) and high ($1000\ s^{-1}$) strain rate behavior of AA7075-T6 aluminum alloy sheet was characterized at room temperature. The anisotropic response of this metal was characterized in terms of differences in true stress response and Lankford coefficient or r-values (the ratio of transverse to thickness strain during uniaxial tensile straining) [8] measured along the rolling (0°), diagonal (45°) and transverse (90°) directions of the sheet. Shear tests were also performed at $0.001\ s^{-1}$ strain rate to characterize anisotropy, as well as to determine the flow stress curves for larger strains. Finally, the Hockett–Sherby [9] constitutive model and Barlat YLD2000 [10] yield criterion were calibrated to the measured data sets.

T. Rahmaan (✉) • C. Butcher • M.J. Worswick
University of Waterloo, Waterloo, ON, Canada
e-mail: trahmaan@uwaterloo.ca

Y. Zhu, A.T. Zehnder (eds.), *Experimental and Applied Mechanics, Volume 4*,
Conference Proceedings of the Society for Experimental Mechanics Series, DOI 10.1007/978-3-319-42028-8_14

14.2 Experimental Setup

The material studied was 2 mm AA7075-T6 sheet. Tensile specimens were tested under room temperatures conditions with the tensile axis oriented along the sheet rolling direction (RD), transverse direction (TD, 90° to rolling direction), and diagonal direction (DD, 45° to rolling direction). Shear specimens were tested with the principal strain direction initially aligned with the RD.

Two different specimen geometries were used in this work as shown in Fig. 14.1, a miniature tensile "dog-bone" specimen and a "mini shear" specimen developed by Peirs et al. [11] that was applied to automotive steel and aluminum alloys by Rahmaan et al. [2] and Abedini et al. [12]. The miniature dog-bone specimens were used for tensile experiments, while the mini shear specimens were used for the shear experiments. As shown by Smerd et al. [3], the short gauge length of the miniature dog-bone specimen is required for high rate characterization, while Rahmaan et al. [5] showed that the true stress response and r-values measured using miniature dog-bone tensile specimens for several automotive sheet alloys were equivalent to results from ASTM-E8 sub-size tensile specimens. The reader is referred to [12] for a detailed description of the mini shear specimen fabrication and experimental data analysis method used in this work.

A servo-mechanical MTS criterion model 45 testing frame was used to conduct uniaxial tensile experiments at low (0.001 and 0.01 s^{-1}) uniaxial strain rates and shear experiments at a von Mises equivalent strain rate of 0.001 s^{-1}. The load was measured using a 100 kN load cell. Cross-head velocities of 0.0125 and 0.125 mm/s were used to achieve nominal strain rates of 0.001 and 0.01 s^{-1}, respectively, in a tensile loading condition, while a cross-head velocity of 0.003 mm/s was used to attain a nominal von Mises equivalent strain rate of 0.001 s^{-1} in shear. The axial and width strains were obtained using 3-D full-field strain measurements using the VIC3D stereo DIC system from Correlated Solutions Inc. The imaging techniques and DIC measurements are the same as those adopted in [12]. A tensile split Hopkinson bar (TSHB) apparatus, described in [13], was used to test the tensile samples at a strain rate of 1000 s^{-1}. In the present work, 15.88 mm diameter 4340 steel round bars with lengths of 2159 and 1829 mm were used for the incident and transmitter bars respectively, while a 508 mm long hollow tube was used as a striker.

14.3 Stress-Strain Response

Three to five repeat tests were performed for each test condition. An "average" stress–strain curve was constructed by interpolating a curve through each experimental data set, obtaining the flow stress at 0.0005 strain increments for each curve and then averaging the values to obtain the average stress value at each strain increment. The average stress values at each strain level were then used to construct the average curve.

The effect of strain rate on the true stress vs. effective plastic strain (or flow stress) along the rolling direction of the AA7075-T6 sheet is shown in Fig. 14.2. The average flow stress at yield and 6 % effective plastic strain is plotted against the strain rate in Fig. 14.3. The error bars in Fig. 14.3 indicate the maximum and minimum measured values from the population of repeat tests. An increase in yield strength with increasing strain rate was observed for both materials in all three sheet orientations. The hardening rate slightly decreases for AA7075-T6 sheet alloy at 1000 s^{-1} strain rate relative to that observed for the lower strain rates (Fig. 14.2). This behaviour may be due to adiabatic heating and thermal softening at the higher strain rate, however, understanding such behavior requires further experimental analysis and therefore was left for future research. The general stress-strain behavior is similar for diagonal and transverse sheet orientations and is not shown for brevity.

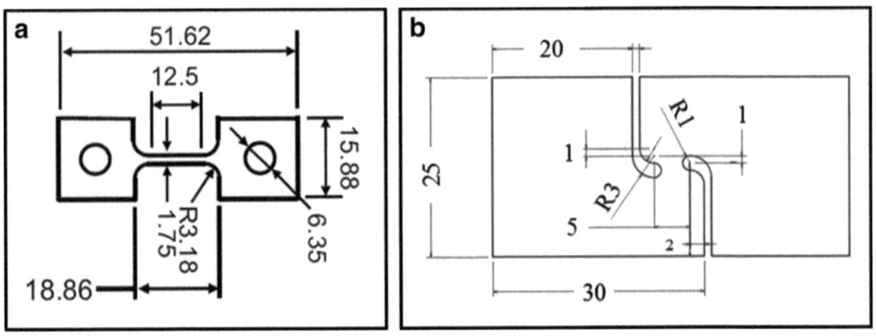

Fig. 14.1 Specimen geometry of (**a**) the miniature tensile specimen and (**b**) the mini-shear specimen [11] (dimensions are in mm)

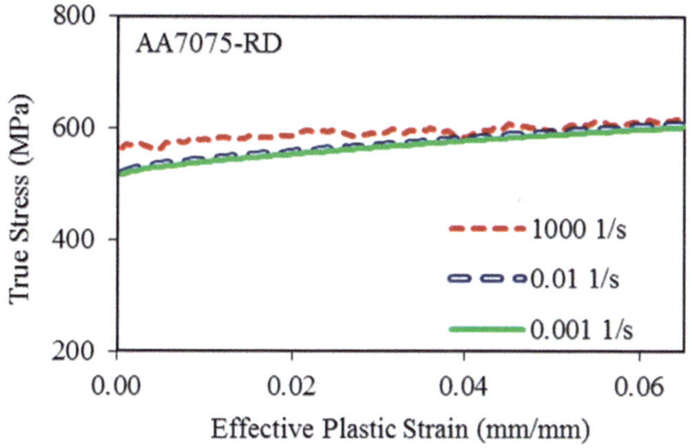

Fig. 14.2 Average flow stress curves of AA7075 sheet specimens at room temperature and strain rate from 0.001 to 1000 s^{-1} along rolling direction

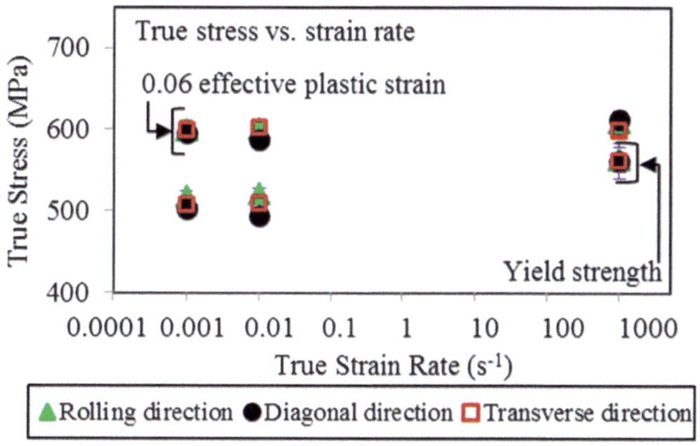

Fig. 14.3 True stress vs. strain rate at yield and 6 % effective plastic strains. The *error bars* indicate the maximum and minimum measured values from the population of repeat tests

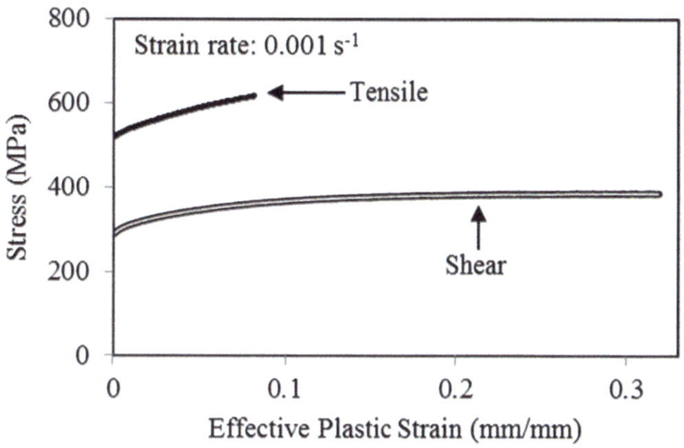

Fig. 14.4 Shear stress-strain curves of AA7075-T6 sheet at room temperature

Figure 14.4 shows the average shear stress vs. effective plastic strain curve for the AA7075-T6 sheet material at 0.001 s^{-1} strain rate. Since the mini-shear geometry does not have an intrinsic length scale, the shear strains were obtained by averaging the measured DIC shear strains in the gauge area as suggested by [2, 12]. Also plotted is the uniaxial tensile data from Fig. 14.2.

14.4 Anisotropy Characterization

For each characterization experiment, the plastic work was computed and applied stress was normalized by the stress in the rolling direction from a tensile test at the same level of plastic work The plastic work, W^P, of the sheet material is determined using Eq. (14.1), is defined:

$$W^P = \int_0^{\varepsilon} \sigma_{ij} d\varepsilon_{ij}^P \tag{14.1}$$

where, σ_{ij} is the stress tensor, and ε_{ij}^P is the plastic strain tensor.

The normalized stresses are plotted against the plastic work in Fig. 14.5 for quasi-static strain rate conditions. The average normalized stresses in the DD and TD uniaxial cases and for shear loading are calculated to be 0.99, 1.00 and 0.58 of the uniaxial RD stress, respectively. This data demonstrates that this alloy displays planar isotropy in terms of the measured uniaxial stress response. Furthermore, the shear stress ratio corresponds to the value of 0. $1/\sqrt{3}$ (0.577) expected for a Mises material.

Based on the plastic work analysis, the measured shear stress-strain data was converted to uniaxial tensile values using von-Mises criterion and assuming planar-isotropy. This conversion was limited to 30 % equivalent strain in order to stay within the infinitesimal strain regime [2]. The converted uniaxial curve (based on the measured shear response) is compared with the measured data obtained from uniaxial tensile testing along the rolling direction at 0.001 s^{-1} (Fig. 14.6). The uniaxial tensile curves generated from the shear data show excellent agreement with the uniaxial tensile test data at low strain rate, supporting the use of the shear experiment to extend the useful strain range of the uniaxial tensile curve. Future work will

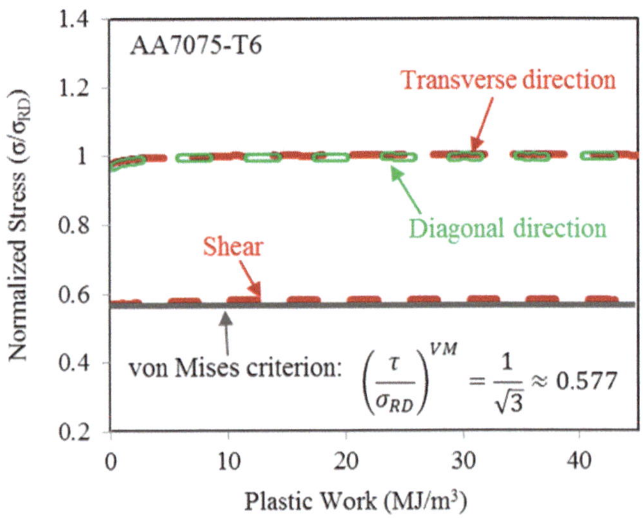

Fig. 14.5 Normalized shear stress-strain curves of AA7075-T6 sheet specimens at room temperature. The *solid line* indicates the expected ratio of shear to uniaxial stress predicted using a von Mises yield criterion

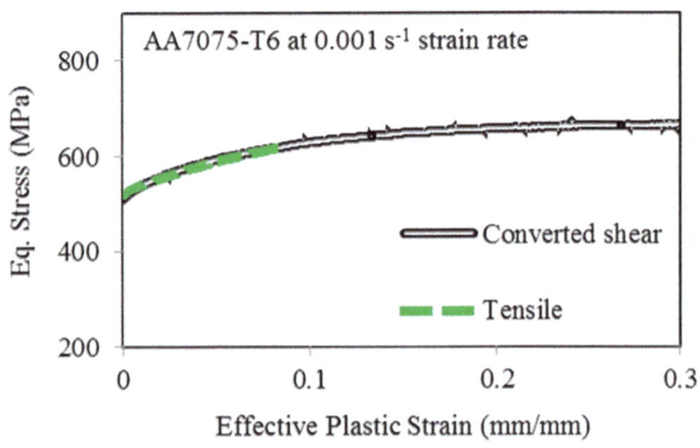

Fig. 14.6 Equivalent stress vs. effective plastic strain comparison for experimental data from tensile and shear testing at 0.001 s^{-1} strain rate

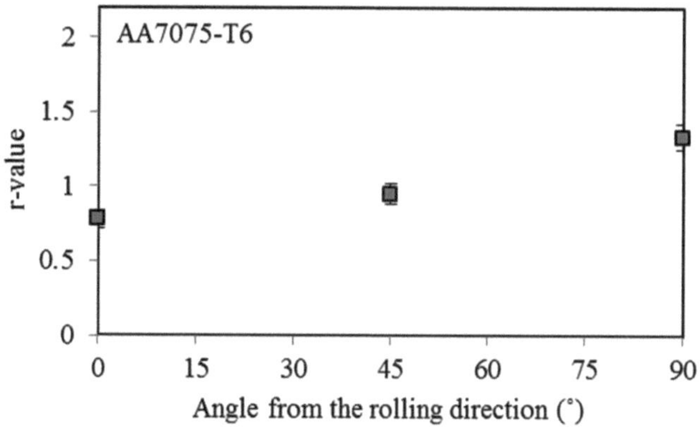

Fig. 14.7 Average Lankford coefficients or r-value for AA7075-T6 sheet 0.001 s^{-1} strain rate

focus on high strain rate shear experiments to extrapolate the tensile flow stress response of these materials at high strain rates and further examine thermal softening effects at high rates.

Although, the AA7075-T6 sheet tested herein exhibited minimal anisotropy in the stress response, this alloy did display appreciable normal and planar anisotropy in terms of the measured r-values (Lankford coefficient). The r-value extracted from a tensile experiment oriented at an angle, θ, relative to the sheet rolling direction is defined by calculating the ratio of the width to thickness true plastic strain, as shown in Eq. (14.2). The DIC strain data provided the true axial and width strain, which were then used to calculate the true thickness strain using the principle of volume conservation and Hooke's law [14].

The r-values for each experiment were computed over a plastic strain range from 0 to 6 % and averaged over each set of repeated tests. The calculated r-values for uniaxial tensile loading in the RD, DD and TD were of 0.78, 0.95, 1.34, respectively. This data is plotted in Fig. 14.7, along with calculated error bars which indicate the maximum and minimum measured values from the population of repeat tests. From this data, the scatter is seen to be low and the r-value is greatest in the TD orientation and the lowest for the RD orientation.

$$r = \frac{\varepsilon^p_{width}}{\varepsilon^p_{thickness}} \tag{14.2}$$

where, ε^p_{width} and $\varepsilon^p_{thickness}$ are the true plastic width and thickness strain, respectively.

14.5 Yield Criterion Calibration

Due to large difference in r-values in different sheet orientations, the Barlat YLD2000 [10] yield criterion was considered in order to capture the plastic anisotropy of the AA7075-T6 sheet. Material data used in the identification process of the Barlat YLD2000 parameters correspond to three uniaxial tensile stresses (σ_0, σ_{45}, σ_{90}), three anisotropic coefficients (r_0, r_{45}, r_{90}), and two simple shear stresses (τ). Simple shear testing in the DD is currently in progress; therefore, the measured simple shear stress in the RD was assumed equal to that in the TD for the yield surface calibration. This approach was deemed reasonable given the observed isotropy of the measured stress response. The values presented in Table 14.1 were used to calibrate the model at plastic work levels of 46 MJ/m^3, corresponding to plastic logarithmic strains of 0.08 in a tensile test in the rolling direction of AA7075-T6.

Figure 14.8 shows the yield surface fit to the AA7075-T6 room temperature data and Table 14.2 shows the corresponding Barlat YLD2000 [10] coefficients. As shown in Fig. 14.9, the predicted data is in good agreement with the experimental data. The error bars in Figs. 14.8 and 14.9 indicate the maximum and minimum measured values from the population of repeat tests. The biaxial stress ratio and r-value were predicted to be 1.00 and 1.01, respectively. Further experimental validation for the biaxial data prediction is in progress.

Table 14.1 Stress ratios and r-values used in the constitutive fits for AA7075-T6 sheet

Alloy	Stress Ratio			r-value		
	σ_{45}/σ_0	σ_{90}/σ_0	τ/σ_0	r_0	r_{45}	r_{90}
AA7075-T6	1.00	1.00	0.58	0.78	0.95	1.34

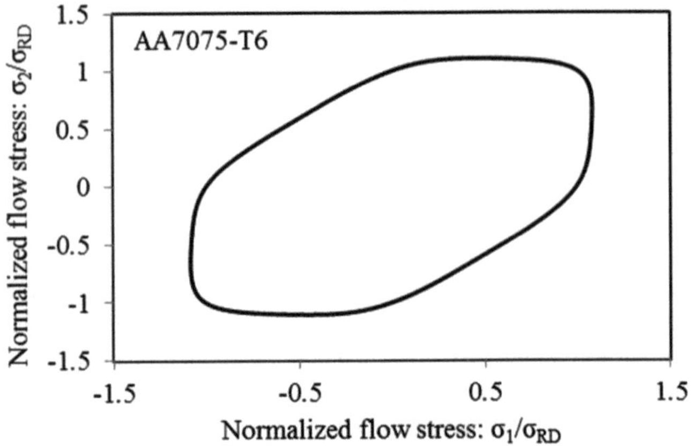

Fig. 14.8 Predicted Barlat YLD2000 yield surface for AA7075-T6 sheet at plastic logarithmic strains of 0.08 in a tensile test in the rolling direction

Table 14.2 Calculated Barlat YLD2000 [10] coefficients for AA7075-T6 sheet at plastic logarithmic strains of 0.08 in a tensile test in the rolling direction

Alloy	Parameters								
	α_1	α_2	α_3	α_4	α_5	α_6	α_7	α_8	m
AA7075-T6	0.89	1.09	1.97	0.42	0.47	1.93	0.99	1.08	8

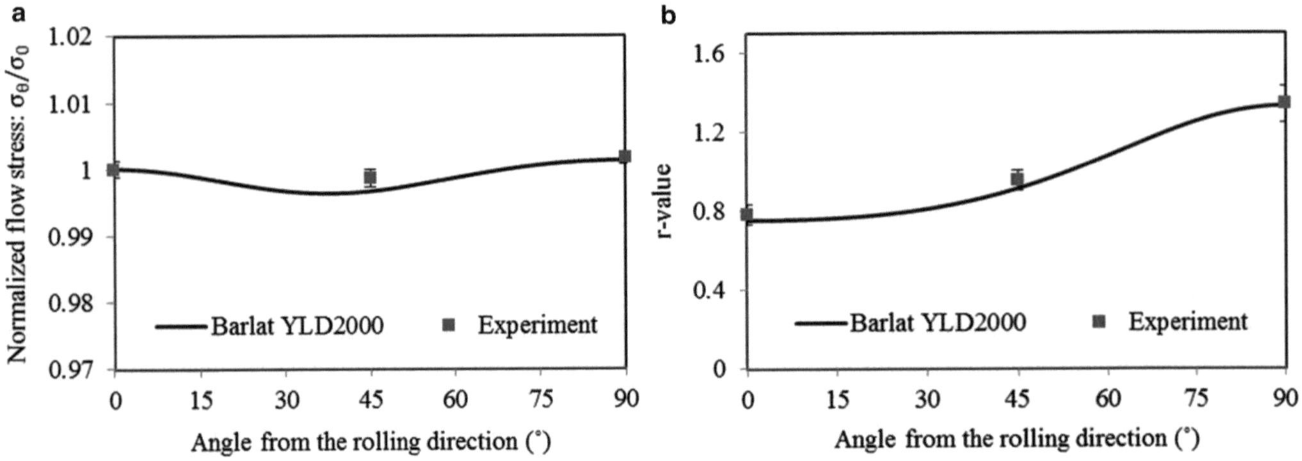

Fig. 14.9 Experimental and predicted (**a**) uniaxial stresses and (**b**) anisotropic coefficients for AA7075-T6 sheet alloy

14.6 Hardening Model Calibration

Constitutive models were fit to the RD quasi-static stress-strain data using non-linear regression analysis. Since the present work is still in progress, strain rate and temperature effects were not accounted for in the models presented in this work. Additional high strain rate experiments in the intermediate range of 10–100 s^{-1} are in progress and the calibration of a strain-rate based constitutive model will be addressed in a future contribution. For the current AA7075-T6 sheet metal alloy, the Hockett–Sherby [9] constitutive model (Eq. 14.3) was fit to the uniaxial stress-strain data converted from the shear

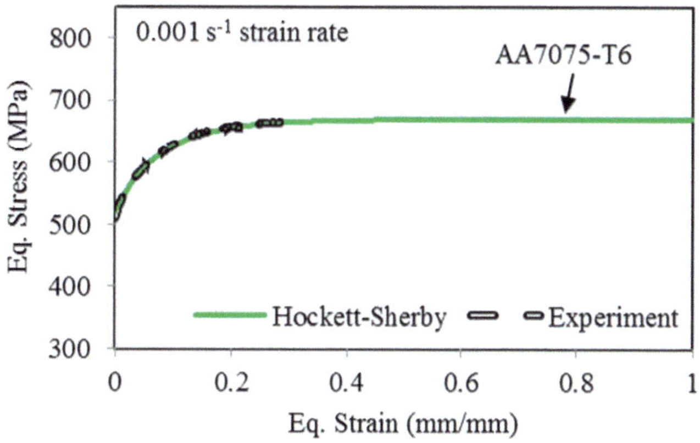

Fig. 14.10 Hockett-Sherby constitutive model fit to the experimental data for AA7075-T6 sheet along rolling direction at 0.001 s^{-1} strain rate

Table 14.3 The Hockett–Sherby constitutive model parameters for AA7075-T6 sheet along rolling direction at 0.001 s^{-1} strain rate

	Parameters			
Alloy	σ_Y (MPa)	σ_S (MPa)	N	P
AA7075-T6	516.87	668.94	11.05	0.929

experiments (Fig. 14.6) since this data it is not limited by the onset of necking and encompasses a larger range of strain than in the tensile experiments. The agreement between the model fit and the experiment data and the extrapolation using the model is shown in Fig. 14.10. The quality of the fit was similar for the diagonal and transverse direction experiments (not shown).

$$\sigma = \sigma_S - (\sigma_S - \sigma_Y)\exp(-N\varepsilon)^P \tag{14.3}$$

where, σ_S is the steady-state flow stress, σ_Y is the yield stress, N and P are material constants. The fitted parameters for the Hockett–Sherby model can be found in Table 14.3.

14.7 Conclusions

The current work has investigated the constitutive behaviour of AA7075-T6 sheet. An anisotropic constitutive model was developed, as well as a fit to the Hockett–Sherby constitutive hardening law at larger strains. Future work will assess the performance of a more complete constitutive model encompassing rate and thermal softening effects.

1. The AA7075-T6 sheet metal alloy tested in this work displays mild strain rate sensitivity;
2. The anisotropy in the stress response was quite low, but significant plastic anisotropy was observed in terms of the measured Lankford coefficients. The average r-values for the RD, DD, and TD orientations were found to be 0.78, 0.95 and 1.34 respectively;
3. The Barlat YLD2000 yield criterion was calibrated to characterize the plastic anisotropy and provides good agreement with the experimental stress ratios and r-values used in its calibration;
4. The Hockett–Sherby constitutive model was able to capture the hardening behavior at 0.001 s^{-1} strain rate.

Acknowledgments The authors greatly appreciate financial support from the Honda R&D Americas, Promatek Research Centre (Cosma International), Alcoa Technical Center, the Natural Sciences and Engineering Research Council of Canada, the Canada Research Chairs Secretariat, and the Ontario Research Fund.

References

1. Salisbury, C., Worswick, M., Mayer, R.: High rate constitutive modeling of aluminum alloy tube. J. Phys. IV **134**(1), 43–48 (2006)
2. Rahmaan, T., Butcher, C., Abedini, A., Worswick, M.: Effect of strain rate on shear properties and fracture characteristics of DP600 and AA5182-O sheet metal alloys. In: EPJ Web of Conferences 94, 01033 (2015). Doi: 10.1051/epjconf/20159401033

3. Smerd, R., Winkler, S., Salibury, C., Worswick, M.J., Lloyd, D., Finn, M.: High strain rate tensile testing of automotive aluminum alloy sheet. Int. J. Impact Eng. **34**(1-4), 541–560 (2005)
4. Hadianfard, M.J., Smerd, R., Winkler, S., Worswick, M.: Effect of strain rate on mechanical properties and failure mechanism of structural Al-Mg alloys. Mater. Sci. Eng. A **492**, 283–292 (2008)
5. Rahmaan, T., Bardelcik, A., Imbert, J., Butcher, C., Worswick, M.J.: Effect of strain rate on flow stress and anisotropy of DP600, TRIP780, and AA5182-O sheet metal alloys. Int. J. Impact Eng. **88**, 72–90 (2016)
6. Nicholas, T.: Tensile testing of materials at high rates of strain. Exp. Mech. **21**(5), 177–185 (1981)
7. El-Magd, E., Abouridouane, M.: Characterization, modelling and simulation of deformation and fracture behaviour of the light-weight wrought alloys under high strain rate loading. Int. J. Impact Eng. **32**(5), 741–758 (2006)
8. Ramos, G.C., Stout, M., Bolmaro, R.E., Signorelli, J.W., Turner, P.: Study of a drawing-quality sheet steel I: stress/strain behaviors and Lankford coefficients by experiments and micromechanical simulations. Int. J. Solids Struct. **47**(17), 2285–2293 (2010)
9. Hockett, J.E., Sherby, O.D.: Large strain deformation of polycrystalline metals at low homologous temperatures. J. Mech. Phys. Solids **23**(2), 87–98 (1975)
10. Barlat, F., Brem, J.C., Yoon, J.W., Chung, K., Dick, R.E., Lege, D.J., Pourboghrat, F., Choi, S.H., Chu, E.: Plane stress yield function for aluminum alloy sheets—part 1: theory. Int. J. Plast. **19**(9), 1297–1319 (2003)
11. Peirs, J., Verleysen, P., Van Paepegem, W., Degrieck, J.: Determining the stressestrain behaviour at large strains from high strain rate tensile and shear experiments. Int. J. Impact Eng. **32**, 406–415 (2011)
12. Abedini, A., Butcher, C., Anderson, D., Worswick, M.J.: Fracture characterization of automotive alloys in shear loading. In: SAE World Congress, Paper # 2015-15H-0274 (2015)
13. Rahmaan T: Low to high strain rate characterization of DP600, TRIP780, AA5182-O. Master's thesis, University of Waterloo, Waterloo (2015)
14. Nasser, A., Yadav, A., Pathak, P., Altan, T.: Determination of the flow stress of five AHSS sheet materials (DP600, DP780, DP780-CR, DP780-HY and TRIP780) using the uniaxial tensile and the biaxial Viscous Pressure Bulge (VPB) tests. J. Mater. Process. Technol. **210**(3), 429–436 (2010)

Chapter 15
Characterization and Modeling of Polymeric Foam Under Multi-Axial Static and Dynamic Loading

I.M. Daniel, J.S. Fenner, B.T. Werner, and J.-M. Cho

Abstract A polymeric foam commonly used in composite sandwich structures was characterized under multi-axial loading at strain rates varying from quasi-static to dynamic. Tests were conducted under uniaxial compression, tension, pure shear and combinations of normal and shear stresses. Quasi-static and intermediate strain rate tests were conducted in a servo-hydraulic testing machine. High strain rate tests were conducted using a split Hopkinson pressure bar (Kolsky bar) system made of polycarbonate bars having an impedance compatible to that of the foam material. The typical compressive stress-strain behavior of the polymeric foam exhibits a linear elastic region up to a yield point, a nonlinear elastic-plastic region up to an initial peak or "critical stress" corresponding to collapse initiation of the cells, followed by strain softening up to a local minimum (plateau or saddle point stress) and finally, a strain hardening region up to densification of the foam. The characteristic stresses of the stress-strain behavior vary linearly with the logarithm of strain rate. A general three-dimensional elastic-viscoplastic model, formulated in strain space, was proposed. The model expresses the multi-axial state of stress in terms of an effective stress, incorporates strain rate effects and includes the large deformation region. Stress-strain curves obtained under multi-axial loading at different strain rates were used to develop and validate the elastic-viscoplastic constitutive model. Excellent agreement was shown between model predictions and experimental results.

Keywords Polymeric foams • Dynamic testing • Strain rate dependence • Constitutive modeling

15.1 Introduction

In many applications composite sandwich structures with polymeric foam cores are exposed to high energy and high velocity dynamic loadings producing multi-axial dynamic states of stress. They are commonly used as core materials in lightweight sandwich structures with high strength, stiffness and impact damage tolerance. The mechanical behavior of cellular foams has been investigated and extensively discussed in the literature [1–5]. However, characterization has been in general limited, and few of the models can capture all the characteristic features of structural foams subjected to multi-axial dynamic loads. Some foam materials, such as PVC foams (especially higher density ones), are strain rate dependent, anisotropic, and elastic/viscoplastic materials. Their deformation history during dynamic loading affects critically the integrity of the sandwich structure. Characterization and modeling of their behavior is especially challenging. Several studies have been reported on dynamic characterization of foams and the effect of strain rate [6–13]. Daniel and Rao characterized isotropic and anisotropic foams at strain rates up to 800 s^{-1} [8]. They used a Hopkinson bar system with polymeric (polycarbonate) rods for the high strain rate tests. Ouellet et al. conducted tests over a wider range of strain rates (0.0087–2500 s^{-1}) and modeled the strain rate behavior [11]. More recent studies by Tagarielli et al. [12], Lee et al. [13] and Daniel and Cho [14] also covered a wide range of strains and strain rates. Constitutive modeling over the wide range of strains and strain rates experienced by foams has lagged because of the finite deformations and the anisotropy involved in some foams, with few works reported in the literature [15–17]. This study expands previous ones to dynamic biaxial stress fields.

I.M. Daniel (✉) • J.S. Fenner
Robert McCormick School of Engineering and Applied Science, Northwestern University, 2137 Tech Drive, Evanston, IL 60208, USA
e-mail: imdaniel@northwestern.edu

B.T. Werner
Sandia National Laboratories, Livermore, CA 94550, USA

J.-M. Cho
Hyundai Motor Co., Seoul, South Korea

Y. Zhu, A.T. Zehnder (eds.), *Experimental and Applied Mechanics, Volume 4*,
Conference Proceedings of the Society for Experimental Mechanics Series, DOI 10.1007/978-3-319-42028-8_15

15.2 Material Characterization

The quasi-static characterization of this material was discussed previously by the authors [14]. The material studied was a closed cell PVC foam, Divinycell H250 (DIAB Inc.). The cell microstructure is shown in Fig. 15.1. On a macroscopic scale, the material is orthotropic/transversely isotropic with principal axes as shown in Fig. 15.1. Typical stress-strain curves of this material under tension and compression along principal material directions are shown in Fig. 15.2. A typical foam stress-strain curve exhibits several stages of deformation, a linear elastic region up to the proportional limit (or yield point), a plastic region up to a peak or critical stress, a plateau region followed by "strain hardening" and culminating in global densification. The linear elastic behavior is associated with elastic bending or stretching of the cell walls; the local peak or "critical point" is related to initiation of cell collapse under wall buckling and/or plastic hinge formation; the plateau region is caused by continuing cell collapse and local densification (locked-up); the strain hardening region ends with the final densification stage which corresponds to complete cell collapse and touching of opposite cell walls.

Macroscopic characterization of cellular foams is a challenging problem because of their hyperelastic behavior and tendency for deformation localization due to local collapse of cells and end crushing under compression. Some of the difficulties can be overcome by using stress-controlled and strain-controlled experiments and full-field strain measurement methods [14, 17]. Stress controlled experiments at various orientations with respect to the principal material axes were conducted with coupons of appropriate aspect ratio.

Strain controlled experiments were conducted by using thin plate specimens constrained between metal blocks as shown in Fig. 15.3. Such experiments were conducted at quasi-static and moderate strain rates of 10^{-4} and 1 s^{-1}. The optimum specimen aspect ratio for compressive, shear and combined compressive and shear strain loading was determined by Finite Element Analysis using previously published properties of a different batch of the same material [2]. The higher the aspect ratio the more homogeneous is the state of strain. An aspect ratio of 10 was deemed suitable for compression and shear specimens.

Compressive stress-strain curves for off-axis specimens tested under uniaxial stress are shown in Fig. 15.4 at various orientation with respect to the in-plane direction (1- or 2 -direction). The effect of anisotropy is reflected in the variation in axial modulus and characteristic first peak in the stress-strain curve. The latter, as mentioned before, is the "critical point" of initiation of local collapse of the cell structure.

Compressive stress tests along the in-plane (1) and through-thickness (3) directions were conducted under uniaxial strain at three strain rates, quasi-static (1–5 $\times$ 10^{-4} s^{-1}), intermediate (1–3 s^{-1}) and high (10^3 s^{-1}). Shear stress tests, were conducted with the shear test configuration of Fig. 15.3, at quasi-static and intermediate strain rates. Quasi-static and intermediate rate tests were conducted in a servo-hydraulic testing machine; high rate tests were conducted in a split Hopkinson (Kolsky) Pressure Bar (SHPB) system using polycarbonate bars for better impedance match with the foam

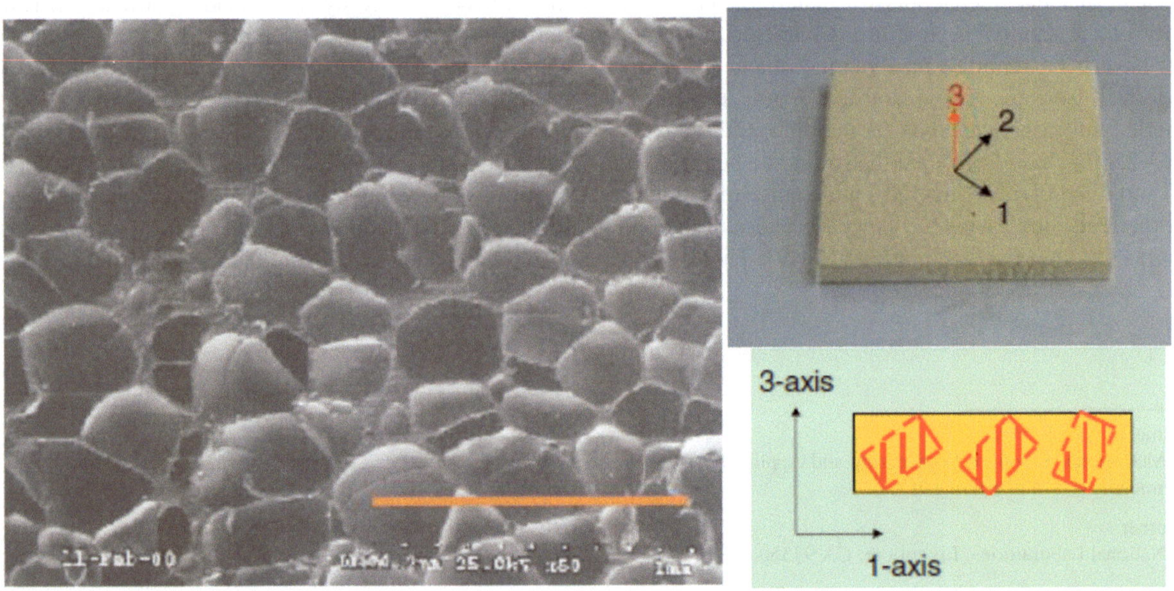

Fig. 15.1 Cell morphology of divinycell H250 foam (scale bar is 1 mm) and principal material axes of foam plate

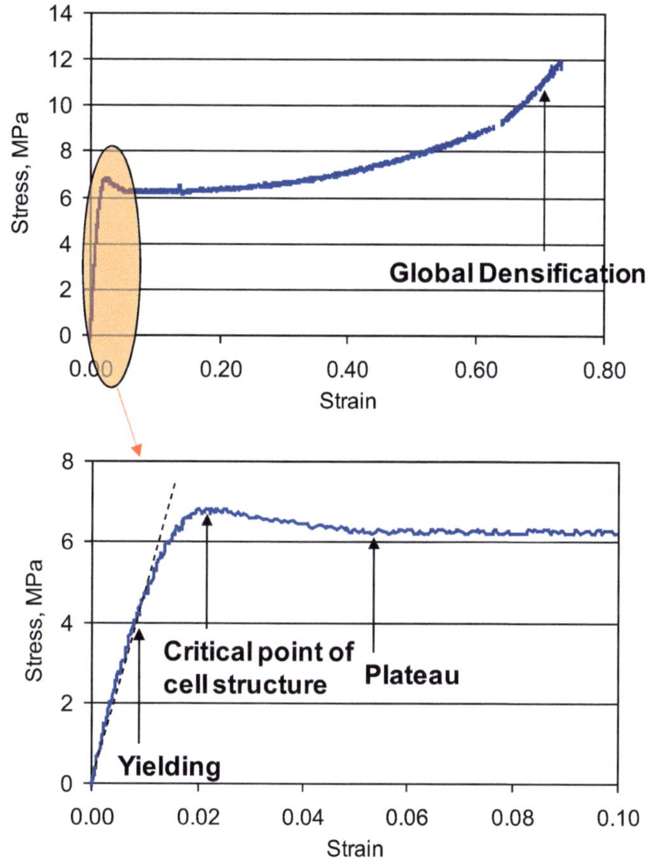

Fig. 15.2 Typical stress-strain curve and characteristic features of a foam under uniaxial compression

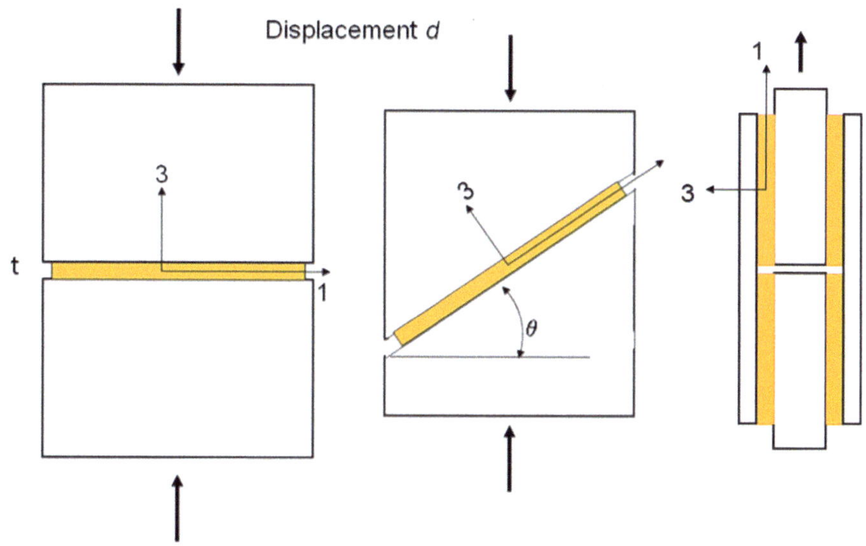

Fig. 15.3 Uniaxial, biaxial and shear strain experiments

material [14, 17]. All strain components other than the one measured were constrained to be zero. These tests yield the stiffness components $C_{11} = C_{22}$, C_{33}, $C_{44} = C_{55}$ and C_{66}.

Biaxial strain tests were conducted by the fixture shown in Fig. 15.3 under quasi-static and intermediate strain rate loadings. In these tests only the vertical load and crosshead displacement were recorded during the test. The strain components can be resolved from purely geometrical considerations as shown in Fig. 15.5. However, the stress components are more difficult to resolve because of the indeterminacy caused by the constraint on the loaded faces of the fixture and the unknown horizontal force components at these faces.

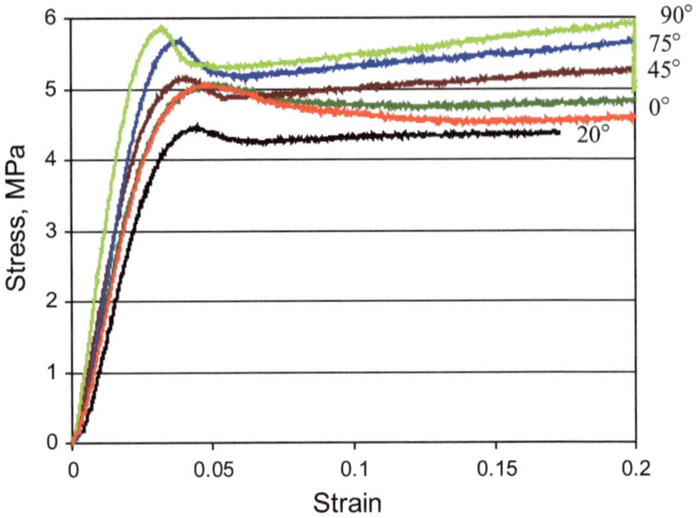

Fig. 15.4 Compressive stress-strain curves of off-axis specimens for loading orientations of 0, 20, 45, 70, and 90° with the 1–2 plane

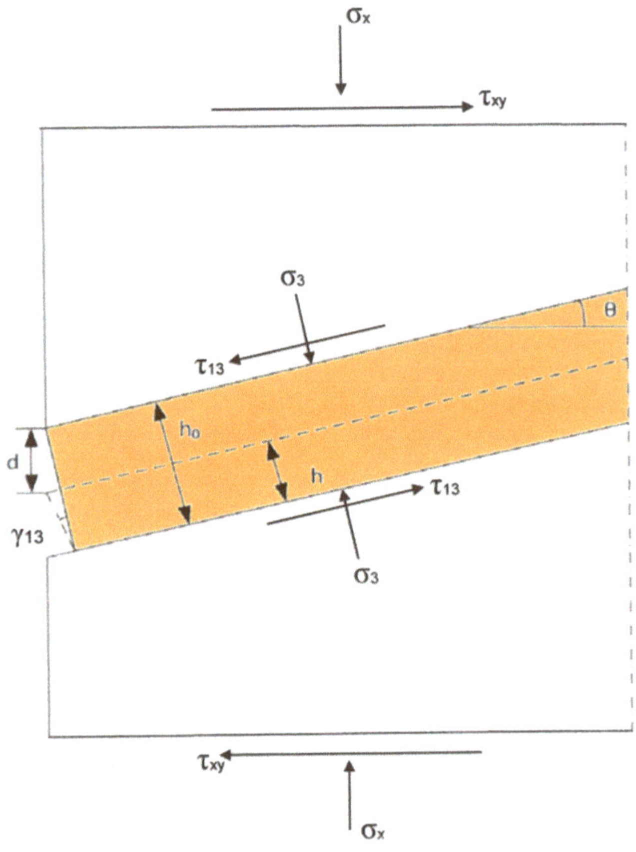

Fig. 15.5 Stresses and strains generated in biaxially loaded under vertical displacement control specimen

The engineering strains are

$$\varepsilon_3 = \frac{d}{h_o}\cos\theta, \ \ \varepsilon_1 = \varepsilon_2 = 0 \tag{15.1}$$

$$\gamma_{13} = \tan^{-1}\frac{d\sin\theta}{h_o - d\cos\theta} = \tan^{-1}\frac{\varepsilon_3\tan\theta}{1-\varepsilon_3}, \quad \gamma_{12} = \gamma_{23} = 0, \tag{15.2}$$

The stresses are expressed as follows

$$\sigma_3 = \sigma_x - \tau_{13}\tan\theta = \sigma_x\cos^2\theta + \tau_{xy}\sin\theta\cos\theta \tag{15.3}$$

$$\tau_{13} = \sigma_3\tan\theta - \tau_{xy} = \sigma_x\sin\theta\cos\theta - \tau_{xy}\cos^2\theta \tag{15.4}$$

The experiments described before provide σ_x, θ, ε_3 and γ_3 and the experimental stress-strain relation

$$\tau_{13} = f(\gamma_{13}) \tag{15.5}$$

15.3 Strain-Rate-Dependent Stress-Strain Behavior

Stress-strain curves obtained at strain rates ranging from 5×10^{-4} to $2 \times 10^3\ s^{-1}$ under nearly homogeneous uniaxial compressive strain in the in-plane and through-thickness directions (1- and 3- directions, respectively) are shown in Figs. 15.6 and 15.7. The stiffness, based on the initial slope of these stress-strain curves, does not vary with strain rate. Four characteristic properties were identified in these stress-strain curves: (1) Yield stress, (2) Peak or "critical" stress corresponding to collapse initiation of the cells, (3) Plateau stress following the initial collapse of the cells, and (4) Strain hardening stress at the end of the plateau region and before the onset of densification. All of the above characteristic stresses increase with strain rate along radial lines from the origin. This behavior is similar to that observed for the polymer matrix of composites [18]. It was also found that these characteristic stresses vary linearly with the logarithm of strain rate. They can be normalized by their values at the reference strain rate ($10^{-4}\ s^{-1}$ in this case) as shown in Fig. 15.8 for the case of in-plane loading. This allows for any property at any train rate to be related to the corresponding property at the reference strain rate as follows:

$$P(\dot{\varepsilon}) = P(\dot{\varepsilon}_o)\left(m\log_{10}\frac{\dot{\varepsilon}}{\dot{\varepsilon}_o} + 1\right) \tag{15.6}$$

where

P = property (yield stress, critical stress, plateau stress, strain hardening stress)

m = 0.104

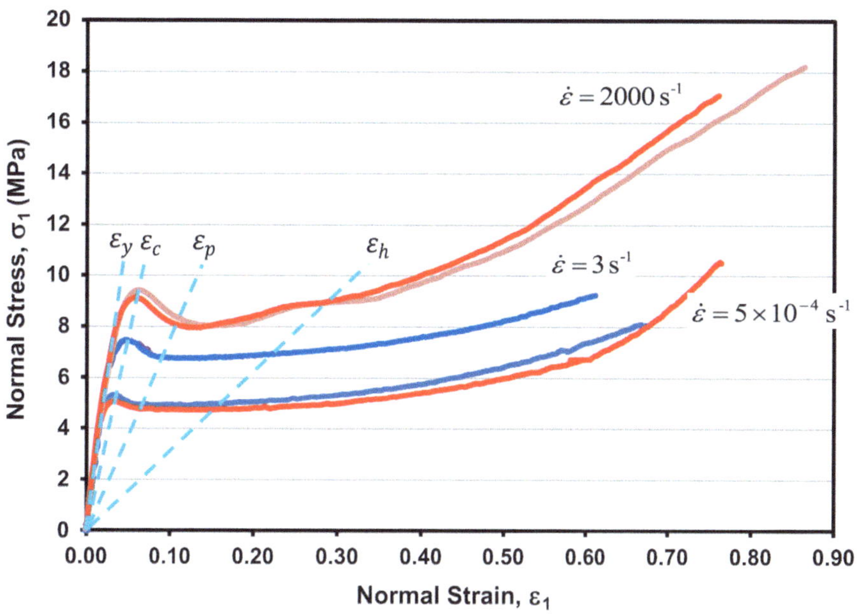

Fig. 15.6 Stress-strain curves at three strain rates (uniaxial strain along in-plane direction)

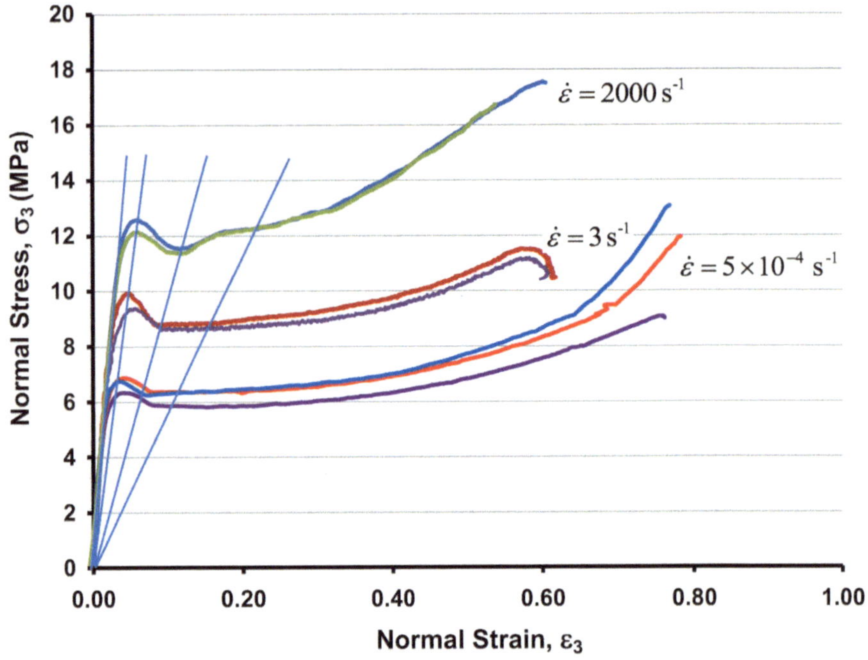

Fig. 15.7 Stress-strain curves at three strain rates (uniaxial strain in through-thickness direction)

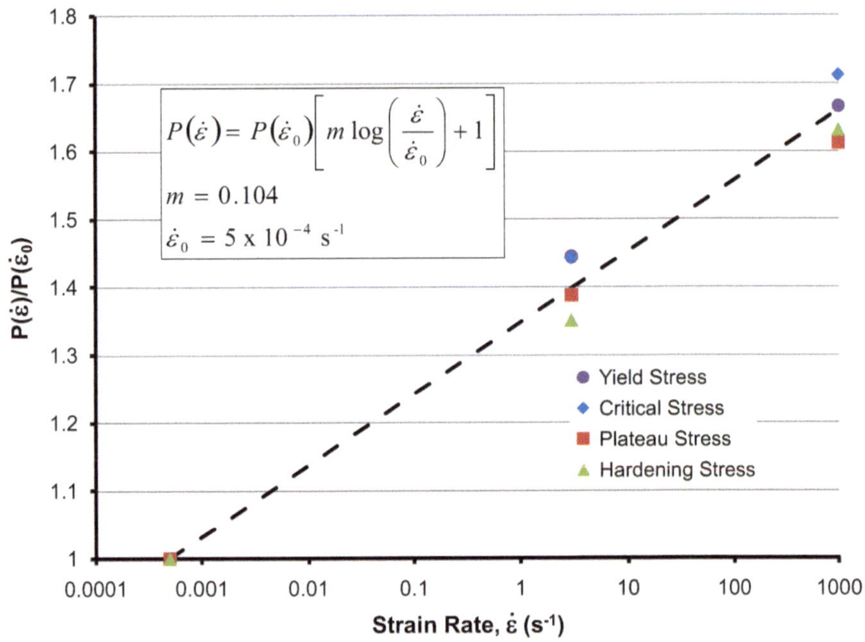

Fig. 15.8 Variation of normalized characteristic properties with strain rate for in-plane compressive loading

$\dot{\varepsilon}_o$ = reference strain rate ($10^{-4}\ s^{-1}$)

Using the above relation, the stress and strain at any strain rate can be transformed into equivalent values at the reference strain rate by the following transformation relations

$$\sigma^*(\dot{\varepsilon}_0) = \frac{\sigma(\dot{\varepsilon})}{K} \qquad \varepsilon^*(\dot{\varepsilon}_0) = \frac{\varepsilon(\dot{\varepsilon})}{K} \tag{15.7}$$

where $K = m\log\left(\frac{\dot{\varepsilon}}{\dot{\varepsilon}_o}\right) + 1$ and the stress-strain curves of Figs. 15.6 and 15.7 can be collapsed into the master curves of Figs. 15.9 and 15.10.

Stress-strain curves at two strain rates under in-plane and out-of-plane shear are shown in Figs. 15.11 and 15.12. Again, the yield stress and critical stress (as defined in Fig. 15.11) vary along radial lines through the origin linearly with the

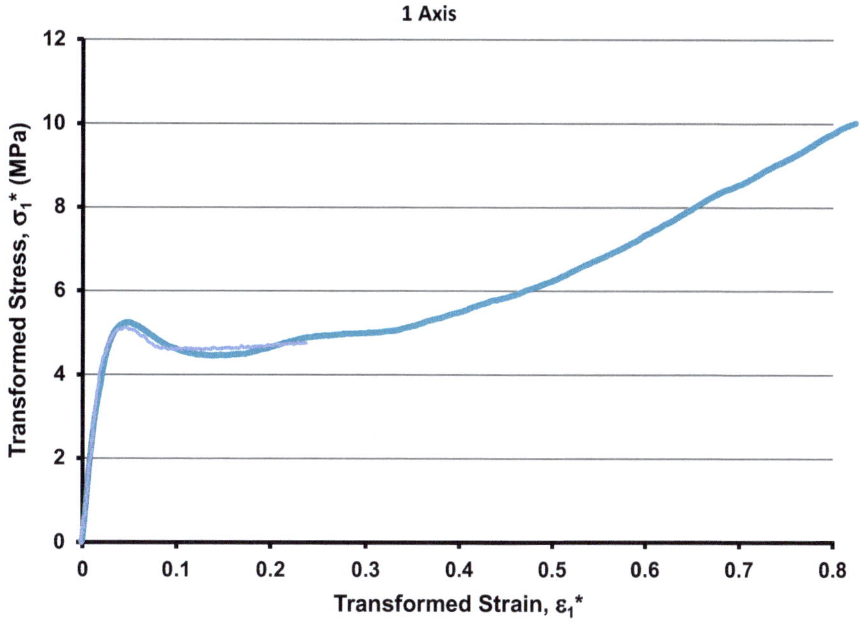

Fig. 15.9 Master stress-strain curve for in-plane compressive loading at any strain rate

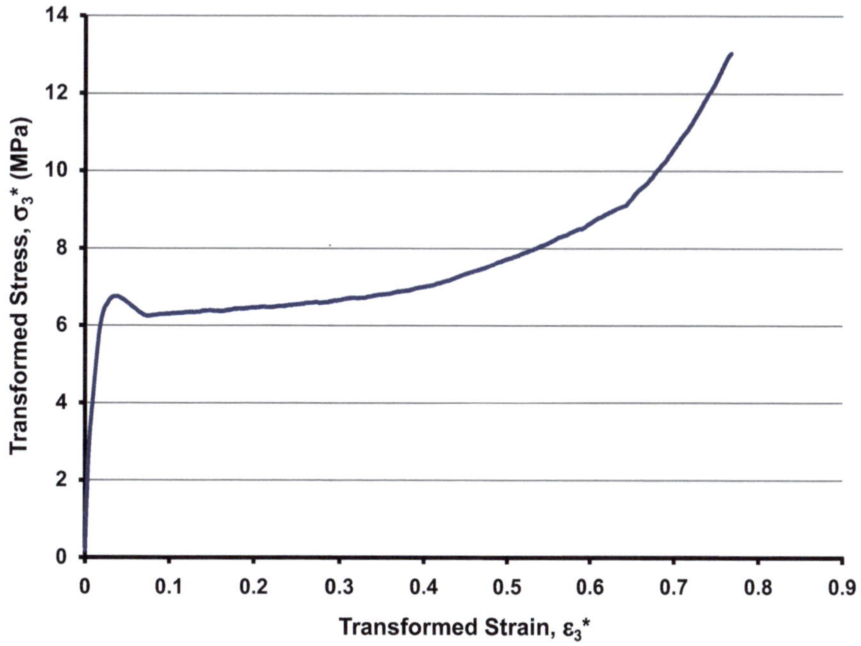

Fig. 15.10 Master stress-strain curve for through-thickness compressive loading at any strain rate

logarithm of strain rate. This allowed the representation of such curves by master stress-strain curves at the reference strain rate as shown if Fig. 15.13.

One characteristic feature in all stress-strain curves is the critical stress which was found to vary linearly with the logarithm of strain rate. This property for the various loadings and strain rates was normalized by its respective quasi-static (reference) value and plotted versus the logarithm of strain rate. It was seen that the normalized critical stress follows the same strain rate law for different loadings normal or shear, in-plane or through the thickness of the anisotropic foam material as was shown for the in-plane loading in Fig. 15.8.

Stress-strain curves at two strain rates, obtained from biaxial tests using the loading scheme of Fig. 15.3, are shown in Figs. 15.14, 15.15 and 15.16. They represent combinations of compressive normal stress and out-of-plane shear stress. In the case of 20° *loading, the behavior is dominated by the transverse compressive stress and the curves, for normal or shear* stress, show a pronounced critical peak. The curves for the 45° loading are equally influenced by the normal compressive and

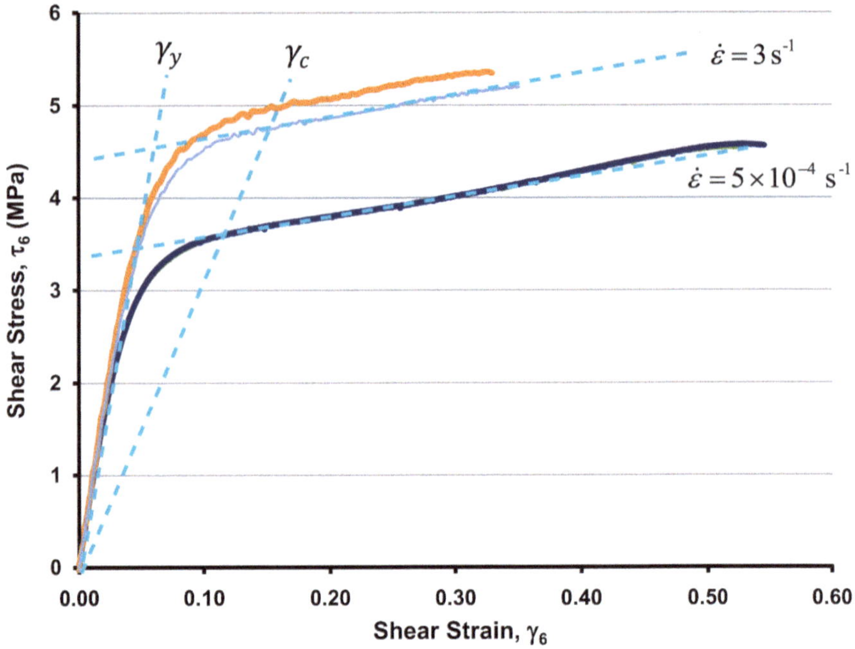

Fig. 15.11 Shear stress-strain curves at two strain rates (in-plane shear)

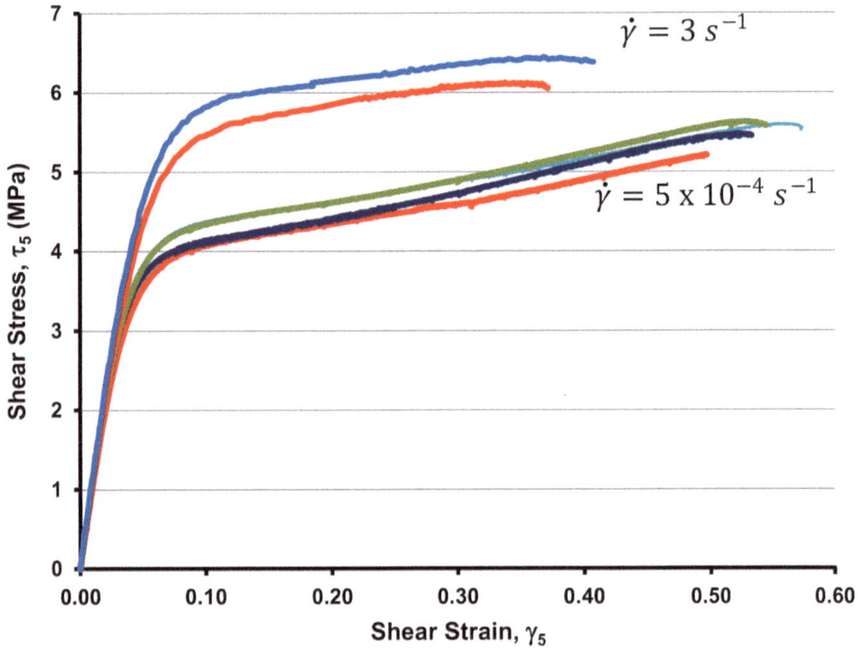

Fig. 15.12 Shear stress-strain curves at two strain rates (through-thickness shear)

through-thickness shear stresses. The critical peaks are less pronounced. The case of 70° loading is dominated by the through-thickness shear behavior. The critical peaks are not distinct, as in the case of pure shear loading, because of the smooth transition from the elastoplastic regime to the strain hardening one. All critical stresses, after normalization by the quasi-static values, vary with strain rate like those discussed before for uniaxial loadings.

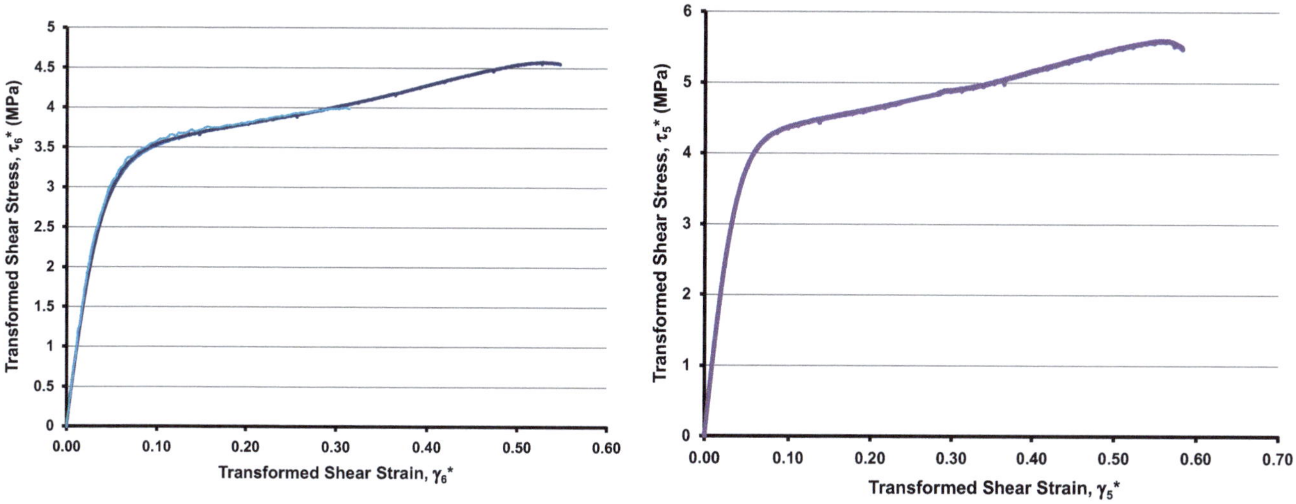

Fig. 15.13 Master shear stress-strain curves for in-plane shear (*left*) and through-thickness shear (*right*)

Fig. 15.14 Biaxial stress-strain curves obtained by loading at 20° with 3-direction

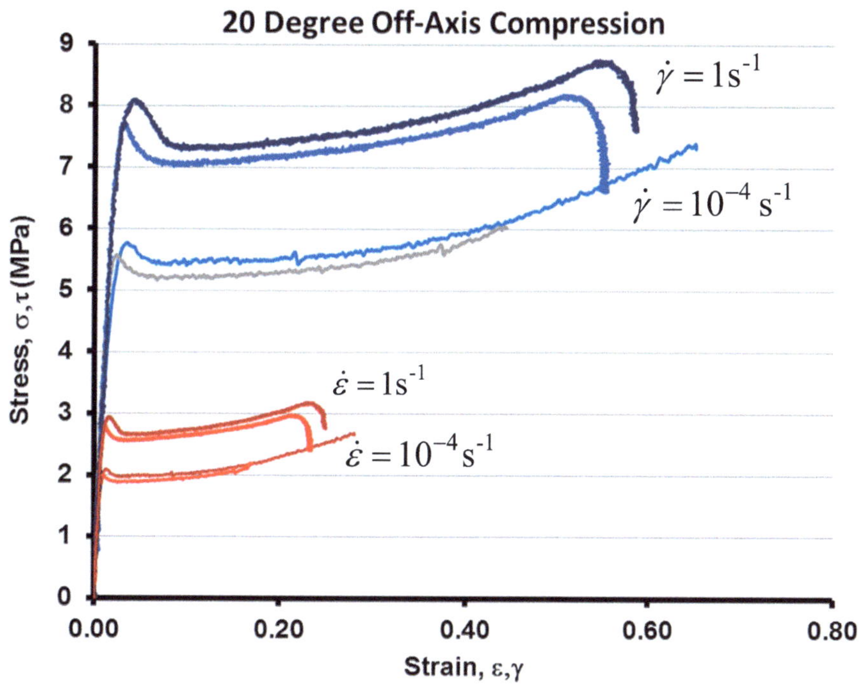

15.4 Constitutive Modeling

In the development above it was shown that the strain-rate dependence of the material behavior can be characterized by one parameter only, the slope m of the characteristic property versus logarithm of strain rate curve. Thus, modeling of the constitutive behavior at one strain rate (reference rate) would be sufficient to describe the mechanical behavior under any state of stress (or strain) and at any strain rate. The following loading function in strain space is proposed.

$$F = \left\{ \begin{array}{l} a_1(\varepsilon_1 - \varepsilon_2)^2 + a_2(\varepsilon_2 - \varepsilon_3)^2 + a_3(\varepsilon_3 - \varepsilon_1)^2 \\ +a_4(\gamma_4)^2 + a_5(\gamma_5)^2 + a_6(\gamma_6)^2 + a_v(\varepsilon_1 + \varepsilon_2 + \varepsilon_3)^2 \end{array} \right\}^{\frac{1}{2}} + b(\varepsilon_1 + \varepsilon_2 + \varepsilon_3) = \overline{\varepsilon} \tag{15.8}$$

where $F = \overline{\varepsilon}$ is the effective strain and a_1, a_2, a_3, a_4, a_5, a_6, a_v, b are material parameters. The potential (loading) function consists of one part associated with deviatoric deformation including the terms with coefficients a_1, a_2, a_3, a_4, a_5, a_6, a quadratic dilatational term with coefficient a_v and a linear dilatational term with coefficient b.

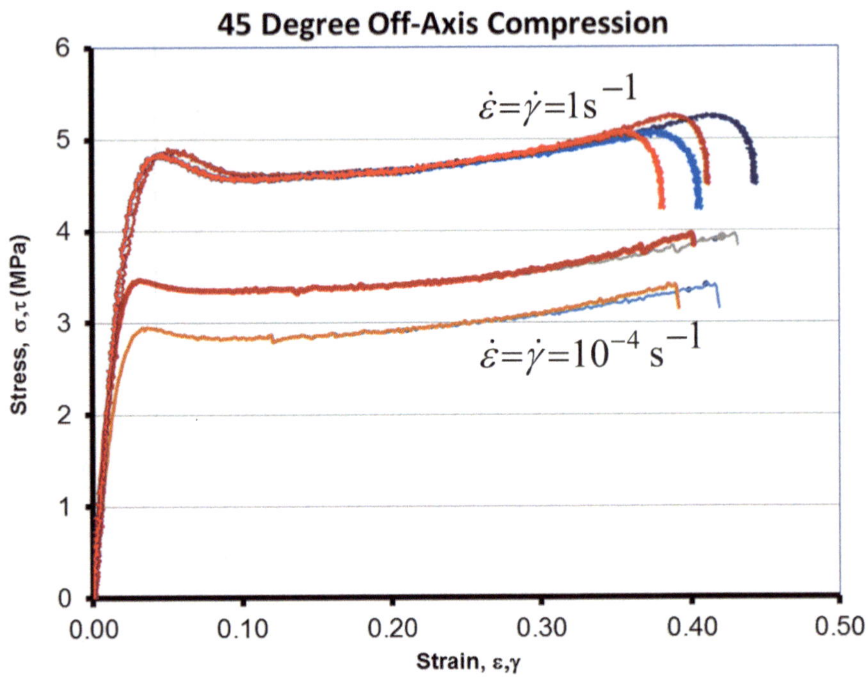

Fig. 15.15 Biaxial stress-strain curves obtained by loading at 45° with 3-direction

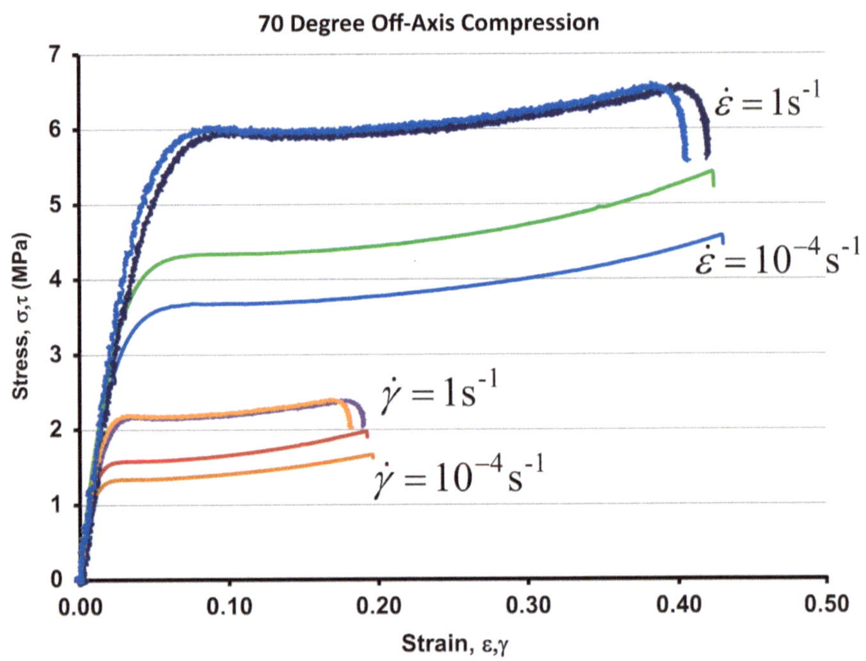

Fig. 15.16 Biaxial stress-strain curves obtained by loading at 70° with 3-direction

For a transversely isotropic material about the 3-axis

$$a_2 = a_3, \qquad a_6 = 2(2a_1 + a_3)$$

All stress-strain curves regardless of stress state or strain rate show similar elastic-plastic behavior up to the critical point. The linear elastic behavior is determined from the characterization tests described before which yielded the C_{ij} stiffness matrix. The plastic behavior has been represented before as a power law [17]. The combination of the linear elastic behavior and the power-law plastic behavior can be expressed in the following incremental constitutive relation

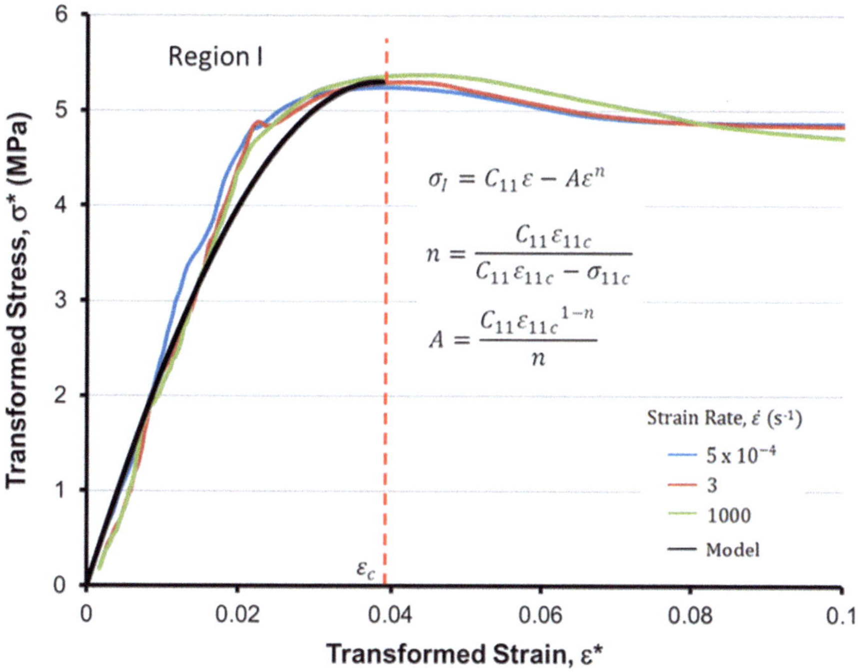

Fig. 15.17 Modeling of elasto-plastic stress-strain behavior for uniaxial loading in 3-direction

$$d\sigma_i = \left[C_{ij} - \left(nA(\bar{\varepsilon})^{n-1}\right)\frac{\partial F}{\partial \varepsilon_i}\frac{\partial F}{\partial \varepsilon_j}\right]d\varepsilon_j \tag{15.9}$$

For example, the stress-strain relation for the case of uniaxial loading in the 3-direction is modeled with a linear term for the linear elastic response and a power law term for the plastic deformation

$$\sigma_3 = C_{33}\varepsilon_3 - A\varepsilon_3{}^n \tag{15.10}$$

where C_{33} is the elastic stiffness and constants A an n are obtained in terms of the critical stress and critical strain [17].

A comparison with experimental results for in-plane loading is shown in Fig. 15.17.

To predict the mechanical behavior of the material for any strain rate and any state of stress, the strain rate parameter m, the elastic constants of the material C_{ij}, the parameters of the potential function, and the power law curve fitting parameters A and n must be determined from experiments. The parameter m is determined from a few controlled tests at different (at least three) strain rates as discussed before. The material stiffnesses C_{ij} can be obtained by quasi-static (at the reference strain rate) characterization. The five independent parameters ..... a_1, a_3, a_5, a_v, b .... can be determined by five independent quasi-static tests, e. g., tests in the 1- and 3-directions, two shear tests and a biaxial test.

15.5 Summary and Conclusions

An orthotropic polymeric cellular foam used in sandwich structures, was characterized under quasi-static and dynamic multiaxial loading conditions. The material investigated was a closed cell PVC foam, (Divinycell H250), having a density of 250 kg/m^3. Quasi-static (5×10^{-4} s^{-1}) and intermediate (3 s^{-1}) strain rate tests were conducted in a servo-hydraulic machine. High strain rate tests (10^3 s^{-1}) were conducted in a split Hopkinson (Kolsky) Pressure Bar (SHPB). This SHPB system was made of polymeric (polycarbonate) bars for closer impedance match with the foam material.

Four characteristic properties were identified in the compressive stress-strain curves: (1) Yield stress, (2) Peak or "critical" stress corresponding to collapse initiation of the cells, (3) Plateau stress following the initial collapse of the cells, and (4) Strain hardening stress at the end of the plateau region and before the onset of densification. All of the above characteristic stresses vary linearly with the logarithm of strain rate. This allows for the development of a unified (master) stress-strain curve referred to one reference strain rate with the strain rate effect defined by a single parameter, the slope of the linear relation between properties and logarithm of strain rate. A constitutive model was proposed to describe the nonlinear multi-axial behavior based on a potential function in the form of a linear combination of deviatoric and dilatational deformation components.

Acknowledgements The work described here was sponsored by the Office of Naval Research (ONR). We are grateful to Dr. Y. D. S. Rajapakse of ONR for his encouragement and cooperation.

References

1. Gibson, L.J., Ashby, M.F.: Cellular solids, 2nd edn. Cambridge University Press, New York (1997)
2. Daniel, I.M., Gdoutos, E.E., Wang, K.A., Abot, J.L.: Failure modes of composite sandwich beams. Int. J. Damage Mech. **11**, 309–334 (2002)
3. Gdoutos, E.E., Daniel, I.M., Wang, K.A.: Failure of cellular foams under multiaxial loading. Compos. Part A **33**, 163–176 (2002)
4. Flores-Johnson, E.A., Li, Q.A.: Degradation of elastic modulus of progressively crushable foams in uniaxial compression. J. Cell. Plast. **44**, 415–434 (2008)
5. Abrate, S.: Criteria for yielding or failure of cellular materials. J. Sandw. Struct. Mater. **10**, 5–51 (2008)
6. Ramon, O., Mintz, J.: Prediction of dynamic properties of plastic foams from constant strain rate measurements. J. Appl. Polym. Sci. **40**(9–10), 1683–1692 (1990)
7. Zhang, J., Lin, Z., Wong, A., Kikuchi, N., Li, V.C., Yee, A.F., Nusholtz, G.S.: Constitutive modeling and material characterization of polymeric foams. J. Eng. Mater. Technol. **119**, 284–291 (1997)
8. Daniel, I.M., Rao, S.: Dynamic mechanical properties and failure mechanisms of PVC foams. In: Dynamic Failure in Composite Materials and Structures, ASME Mechanical Engineering Congress and Exposition 2000; AMD-243, Orlando, 5–10 Nov. 2000. pp. 37–48
9. Avalle, M., Belingardi, G., Montanini, R.: Characterization of polymeric structural foams under compressive impact loading by means of energy-absorption diagram. Int. J. Impact Eng. **25**, 455–472 (2001)
10. Viot, P., Beani, F., Latallade, J.L.: Polymeric foam behavior under dynamic compressive loading. J. Mater Sci. **40**, 5829–5837 (2005)
11. Ouellet, S., Cronin, D., Worswick, M.: Compressive response of polymeric foams under quasi-static, medium and high strain rate conditions. Polym. Test. **25**, 731–743 (2006)
12. Tagarielli, V.L., Deshpande, V.S., Fleck, N.A.: The high strain rate response of PVC foams and end-grain balsa wood. Compos. Part B **39**, 83–91 (2008)
13. Lee, Y.S., Park, N.H., Yoon, H.S.: Dynamic mechanical characteristics of expanded polypropylene foams. J. Cell. Plast. **46**, 43–55 (2010)
14. Daniel, I.M., Cho, J.M.: Characterization of anisotropic polymeric foam under static and dynamic loading. Exp. Mech. **51**(8), 1395–1403 (2011)
15. Tagarielli, V.L., Deshpande, V.S., Fleck, N.A., Chen, C.: A constitutive model for transversely isotropic foams, and its application to the indentation of balsa wood. Int. J. Mech. Sci. **47**, 666–686 (2005)
16. Gielen, A.W.J.: A PVC-foam material model based on a thermodynamically elasto-plastic-damage framework exhibiting failure and crushing. Int. J. Solids Struct. **45**, 1896–1917 (2008)
17. Daniel, I.M., Cho, J.M., Werner, B.T.: Characterization and modeling of strain-rate-dependent behavior of polymeric foams. Compos. Part A **45**, 70–78 (2013)
18. Werner, B.T., Daniel, I.M.: Characterization and modeling of polymeric matrix under static and dynamic loading. Compos. Sci. Technol. **102**, 113–119 (2014)

Chapter 16
Cyclic Loading Experiment for Characterizing Foam Viscoelastic Behavior

Wei-Yang Lu, Matthew Neidigk, and Nicholas Wyatt

Abstract Several open-cell flexible foams, including aged polyurethane foams, were mechanically characterized over a temperature range of −40 to 20 °C. Quasi-static compression was performed to obtain the stress-strain behavior of the foams. The stress-strain relation is nonlinear, but typically there is a small range of linear behavior initially. Compressive cyclic loading at different amplitudes and frequencies of interest (20–60 Hz) were applied to measure foam's hysteresis properties, i.e. stiffness and energy dissipation. The cyclic characterization includes foams with different amount of pre-strains, some are beyond the initial linear range as occurred in many applications.

Keywords Flexible foam • Viscoelasticity • Nonlinear viscoelasticity • Pre-strain • Soft materials

16.1 Introduction

Flexible foams are used in packaging applications for shock and vibration isolation during shipping and transportation. A common scenario is that the foam component deforms slowly to support the weight of the object when packing. During the transporting environment, the package will be subjected to dynamic load, for example from bumpiness of the road. The loading is vibratory in nature and its amplitude is typically low. The loading profile is known or can be measured. Some critical engineering questions about the packaging are: what are the resonance frequencies and how much force is transmitted through the foam to the object? The foam's behavior is known to be very complex. Modeling and analyzing the problem using non-linear viscoelastic analysis is very challenging. To circumvent this difficulty, the foam can be considered to be undergoing a small amplitude vibration superposed onto a larger compression [1]. The analysis can be simplified by approximating the foam behavior to be linear over the small amplitude of strain of interest; therefore, only linear analysis is needed to analyze this small amplitude vibration problem. The approximated linear viscoelastic properties of the pre-stressed foam, however, need to be characterized experimentally before a validated flexible foam model is developed. This paper presents the experiments performed to obtain linear viscoelastic parameters of several flexible foams at various frequencies, temperatures, and pre-strains.

16.2 Foam's Nonlinear Viscoelastic Behavior

Figure 16.1 shows typical compressive stress-strain behaviors of a flexible open-cell foam TF5070-10 [2], which is 10 pcf (pound per cubic foot, lb/ft^3), at different strain rates 0.1, 100, and 200 s^{-1}. The constant strain rate stress-strain curve is generally nonlinear. The loading portion shows typical elastomeric foam behavior with three regimes of initial linear elasticity, elastic-buckling plateau, and densification [3]. Young's modulus E (the initial slope of the curve) and the buckling stress (the boundary separates the linear elastic regime from that of elastic buckling) are clearly rate dependent; the values of E and σ_{el} at 200 s^{-1} are about quintuple and double, respectively, of those at 0.1 s^{-1}. During unloading, it exhibits a strong hysteresis behavior. When the stress returns to zero, the strain lags behind but eventually goes back to zero with little permanent deformation and damage.

W.-Y. Lu (✉)
Sandia National Laboratories, Livermore, CA 94551-0969, USA
e-mail: wlu@sandia.gov

M. Neidigk • N. Wyatt
Sandia National Laboratories, Albuquerque, NM 87123, USA

Y. Zhu, A.T. Zehnder (eds.), *Experimental and Applied Mechanics, Volume 4*,
Conference Proceedings of the Society for Experimental Mechanics Series, DOI 10.1007/978-3-319-42028-8_16

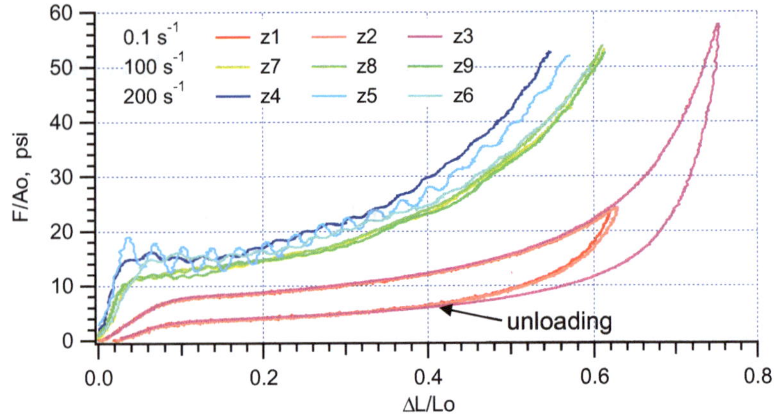

Fig. 16.1 Compressive stress-strain curves of TF5070-10 foam at various strain rates

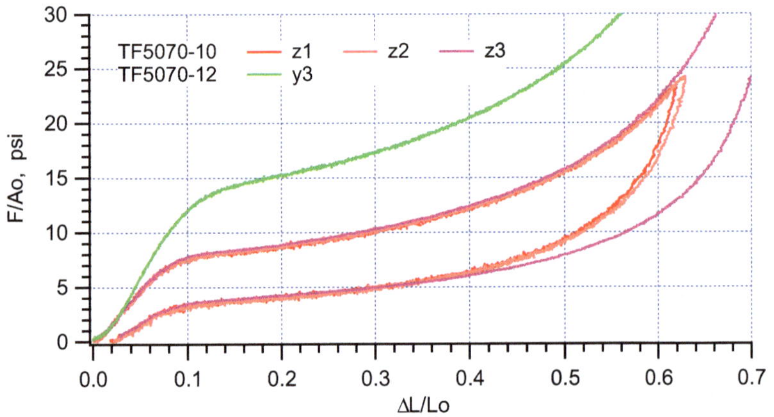

Fig. 16.2 For TF5070 foams, the crush stress increases almost 100 % when density changes from 10 to 12 pcf

Other factors are also known to have significant influences on foam properties such as density and temperature. Figure 16.2 shows the stress-strain curves of foams of two different densities compressed at the same strain rate 0.1 s^{-1}. The yield stress increases almost 100 % when the density changes from 10 to 12 pcf, only a 20 % increase. Figure 16.3 shows the effect of temperature on TF5070-12 (12 pcf) under constant strain rate 0.1 s^{-1}. Notice the vertical axis is in logarithm scale. A flexible foam is generally in rubbery state at the room temperature, where E and σ_{el} change slowly with temperature. When it gets close to the glass transition temperature, a small change in temperature or strain rate may cause those values to change substantially. Modeling the complex nonlinear viscoelastic foam behavior over a wide range of temperature, strain and strain rate is very challenging.

16.3 Material and Specimen

Two foam materials were considered in this study: new and aged 15 pcf open-cell flexible polyurethane foams. The new foam was recently purchased; the aged foam was taken from an old block, which was about 30 years old. Figure 16.4a and b shows scanning electron microscope (SEM) images of these foams. The averaged cell size is about the same 200 μm (0.008 in.). The shape of the cell is slightly different; the new foam's is oval. The foam specimens were cylindrical, the diameter was 10 mm and the length was also 10 mm. The relation between the foam rising direction and the specimen orientation is not known.

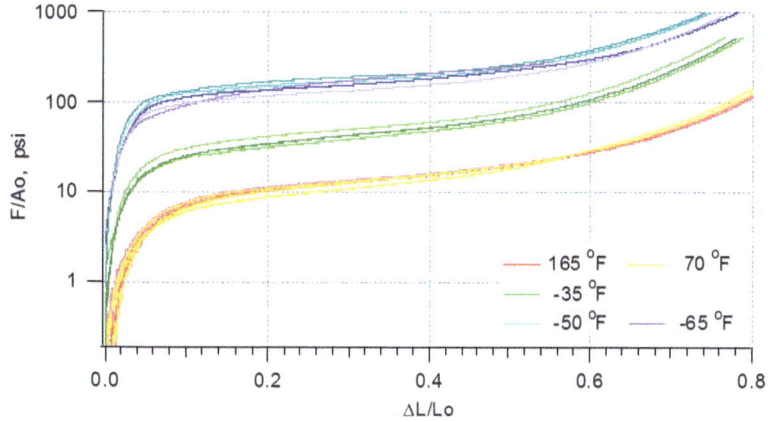

Fig. 16.3 The crush stress could increase more than 10 times at −65°F comparing to the ambient condition

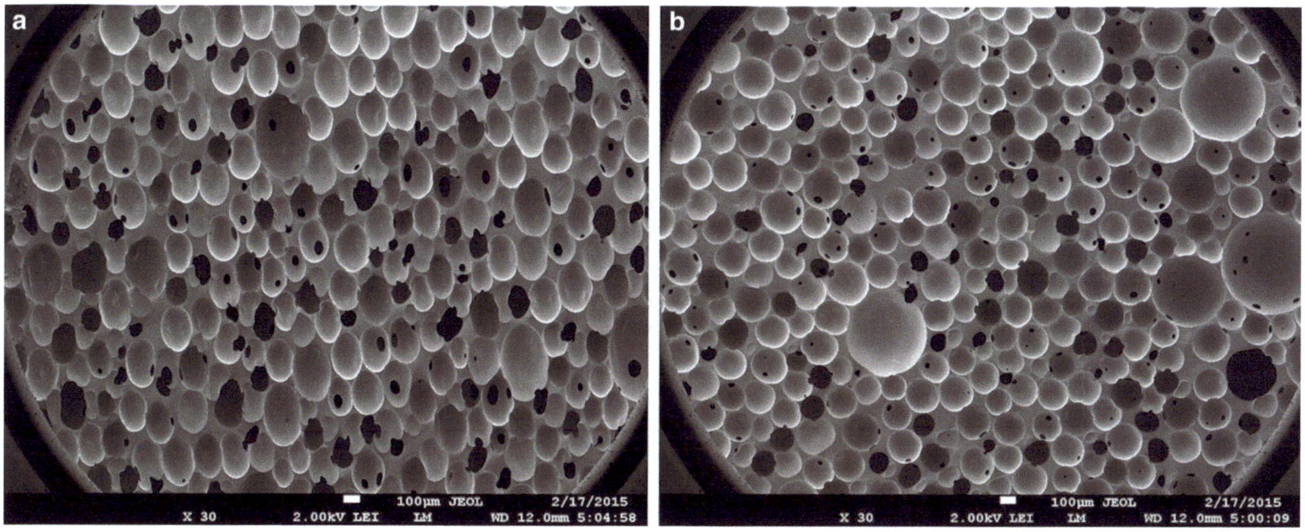

Fig. 16.4 SEM images of 15 pcf foams: (**a**) new and (**b**) aged

16.4 Experimental Setup

The dynamic compression setup is shown in Fig. 16.5. It was based on a Bose Bench Top Testing System with an environmental chamber. The load cell was mounted outside the chamber. Extended rods from actuator and load cell were utilized to position the foam specimen in the chamber.

16.5 Experiment

For this series of compression cyclic test, the intension was to obtain the foam viscoelastic properties within the initial elastic regime. A summary of the tests is listed in Table 16.1. Each specimen was tested under several constant temperature conditions, possibly, 20, 0, −20, −40 °C. At each temperature, the specimen was subjected to three cyclic loading blocks, which were identified by its distinctive loading frequency, in the following order: 20, 40, and 60 Hz. Each loading block contained at least several hundred cycles of deformation, so the load-displacement hysteresis loop was stabilized. All tests were started with the room temperature condition, 20 °C. After completing three blocks of loading, the specimen was cooled down to the next temperature level, a step of −20 °C; and then the three blocks loading was repeated.

Fig. 16.5 Foam dynamic compression experimental setup

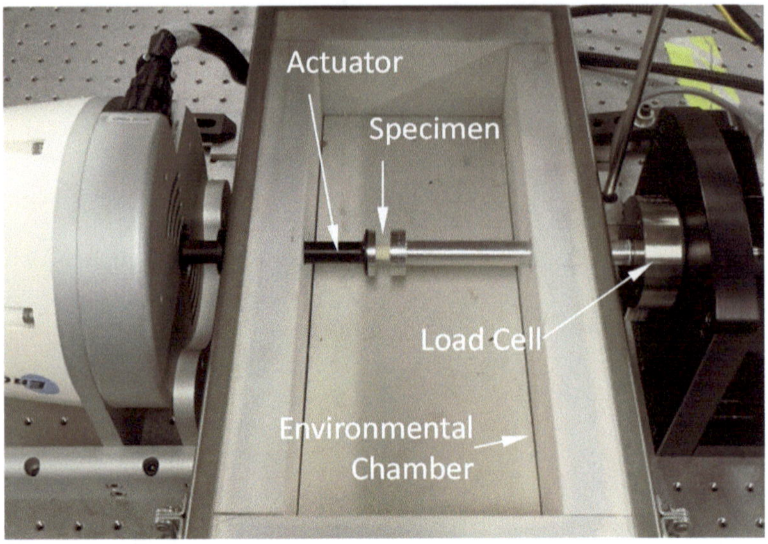

Table 16.1 Summary of foam cyclic experiments

Sample	Temperature, C				Frequency, Hz				Range
	20	0	−20	−40	1	20	40	60	mm
New03	x	x	x			x	x	x	0.25–0.75
New04	x	x	x	x		x	x	x	0.4–0.6
New05	x	x	x	x	x	x	x	x	Vary
Aged05	x	x	x			x	x	x	Vary
Aged06	x	x	x	x		x	x	x	Vary

In general, the data acquisition rate needs to be at least 20 times faster than the cyclic frequency to characterize the dynamic response. Continuous recording would end up with huge amount of unnecessary data. Here, data was acquired only at selected time periods, approximately a period of 0.1 s for every 5 s. The sampling frequency was greater than 1000 Hz.

Figure 16.6a displays the deformation history of specimen New03 at room temperature, where only the recorded data are shown. The initial loading-unloading cycle was done at a strain rate of 0.025 s^{-1} with the purpose of verifying that the boundary of the initial elastic regime is about 10 %. (Detailed compression properties of 15 pcf polyurethane foams are presented in reference [4].) Three blocks of cyclic loading are indicated on the plot, where the displacement of the actuator traveled between −0.50 ± 0.25 mm, corresponding to −5.0 ± 2.5 % strain. The figure does not show cyclic displacement-time curve clearly because of the time scale. Three marked segments are replotted in Fig. 16.6b–d with the extended time scale. The averaged strain rates of 20, 40 and 60 Hz loading were about 2.0, 4.0 and 6.0 s^{-1}, respectively, which were about two orders of magnitude higher than the initial cycle. The force-displacement response is plotted in Fig. 16.7a. The block cyclic loading was all within the initial linear regime and the stable hysteresis loops are quite similar.

The 0 °C response of Specimen New03, with the same pre-strain $\varepsilon_o = -5$ % and cyclic strain amplitude $\Delta\varepsilon = 5$ % as in 20 °C test, is shown in Fig. 16.7b. The hysteresis loop of 0 °C test is clearly larger than that of 20 °C. The 40 Hz hysteresis loop is somewhat distorted at force equals zero, indicating the foam specimen was separated from the platen when displacement was close $d = -0.25$ mm.

An adjustment of the displacement to −0.50 ± 0.10 mm was applied to foam specimen New04. The results are shown in Fig. 16.8. With a smaller displacement amplitude, the specimen was always in contact with the platen during cyclic tests at both 20 and 0 °C. The room temperature results are shown in Fig. 16.8a. The desired pre-strain was $\varepsilon_o = -5.0$ % and the amplitude of strain cycle was $\Delta\varepsilon = 2.0$ % (i.e., ±1.0 %). Figure 16.8b shows the 0 °C results. From the data, the initial loading was delayed after an displacement offset of −0.15 mm, which lead to the pre-strain value of the cyclic test was actually −3.5 %, not −5.0 %. It is because of changing the temperature environment, difficult to know the amount of displacement offset beforehand, and to control the pre-strain accurately.

If the material is linear viscoelastic, dynamic characterization could be done at any pre-strain and strain amplitude and the result will be the same. To characterize the initial viscoelastic regime, the only requirement is that the stress should not exceed the buckling stress and the condition of fixed pre-strain and strain amplitude could be relaxed, which made the

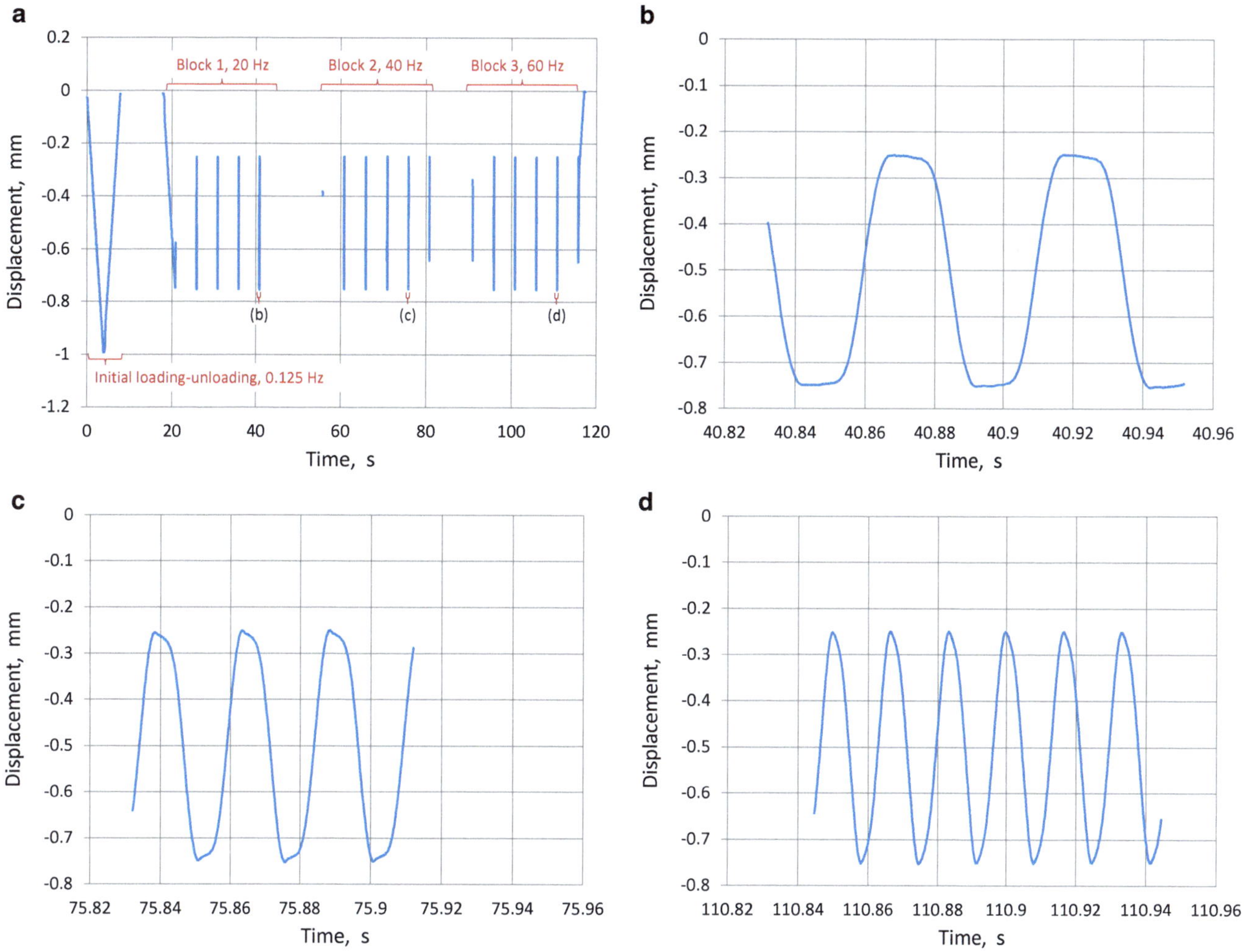

Fig. 16.6 Deformation history of New03 at room temperature

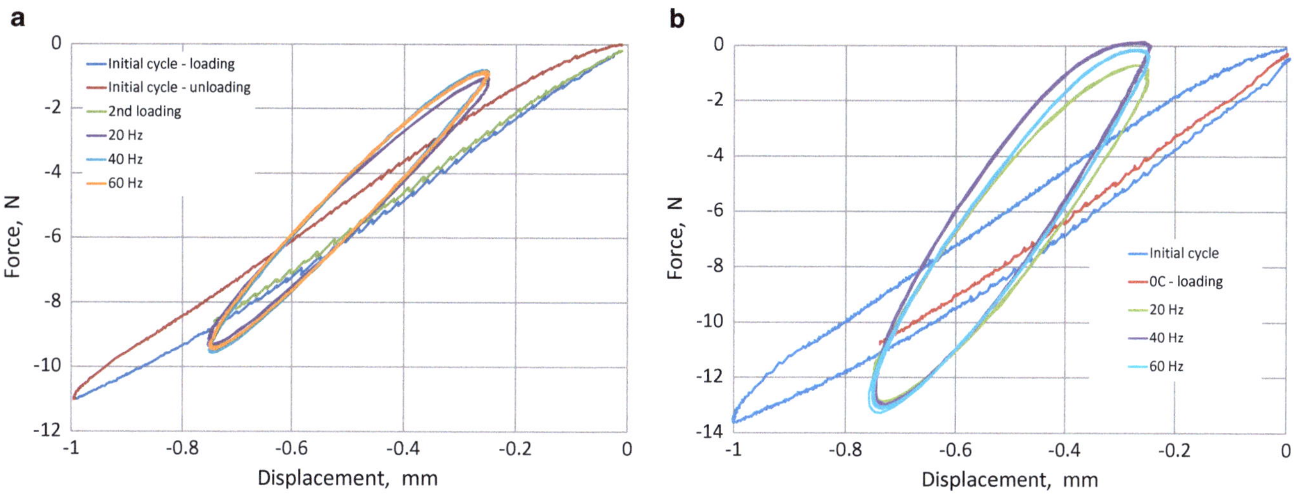

Fig. 16.7 Stable hysteresis loops, −0.50 ± 0.25 mm, of foam specimen New03 at (**a**) 20 °C and (**b**) 0 °C

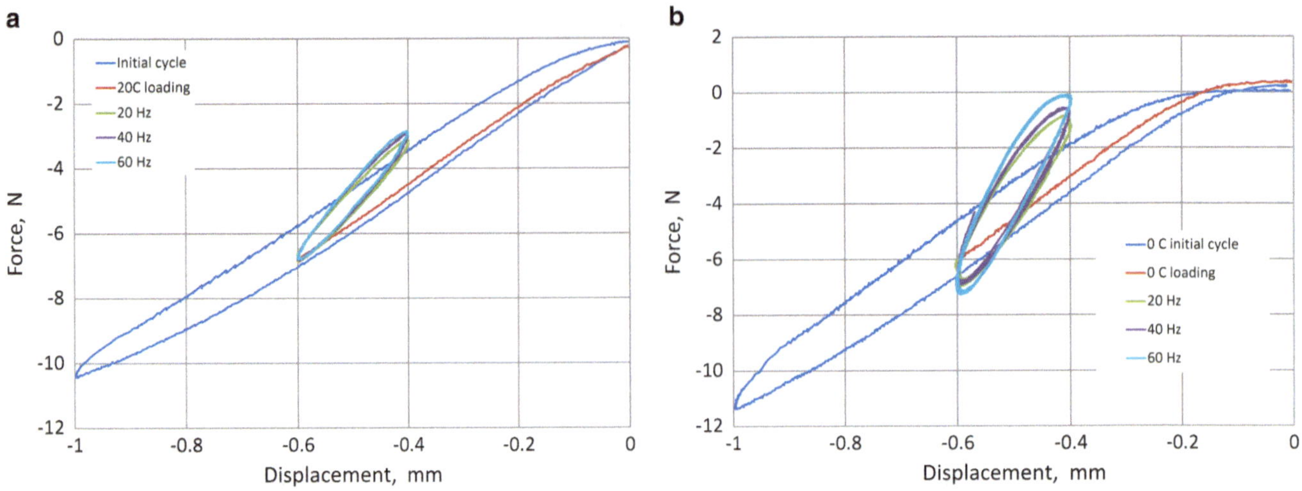

Fig. 16.8 Stable hysteresis loops, -0.50 ± 0.10 mm, of foam specimen New04 at (**a**) 20 °C and (**b**) 0 °C

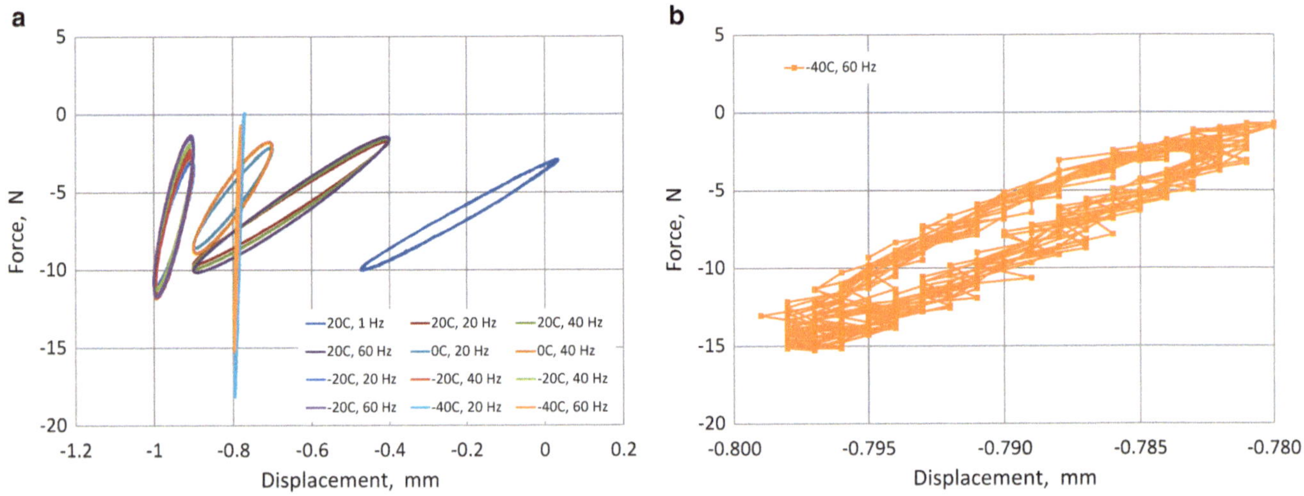

Fig. 16.9 Stable hysteresis loops of new foam specimen New05

characterization experiment easier. Figures 16.9 and 16.10 show the dynamic characterizations of Specimen New05, Aged05, and Aged06. In these tests, both ε_o and $\Delta\varepsilon$ were varied case by case. Also, to compensate the distance change due to different temperatures, both actuator displacement and crosshead motion were applied; therefore, the displacement between cyclic blocks did not represent the actual actuator displacement.

16.6 Viscoelastic Properties of the Initial Linear Regime

For linear viscoelastic material, when cyclic loading reaches equilibrium (i.e., the stress-strain hysteresis loop is stabilized), the stress and strain are sinusoidal but the strain lags behind the stress with a phase difference:

$$\varepsilon = \varepsilon_o + \varepsilon_A \sin(\omega t + \phi_1), \quad \sigma = \sigma_o + \sigma_A \sin(\omega t + \phi_2), \quad \delta = \phi_2 - \phi_1 \tag{16.1}$$

where ω is the angular frequency and δ is the phase lag. The material's storage and loss modulus, *E1* and *E2*, can be calculated by

$$E1 = (\sigma_A/\varepsilon_A)\cos\delta, \quad E2 = (\sigma_A/\varepsilon_A)\sin\delta \tag{16.2}$$

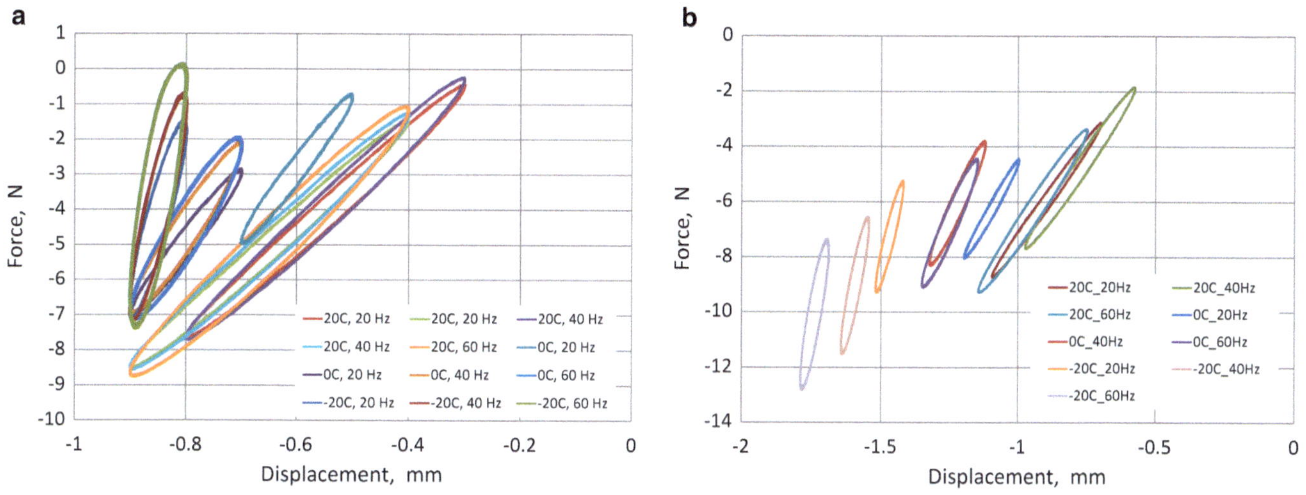

Fig. 16.10 Stable hysteresis loops of aged foam specimen (**a**) Aged05 and (**b**) Aged06

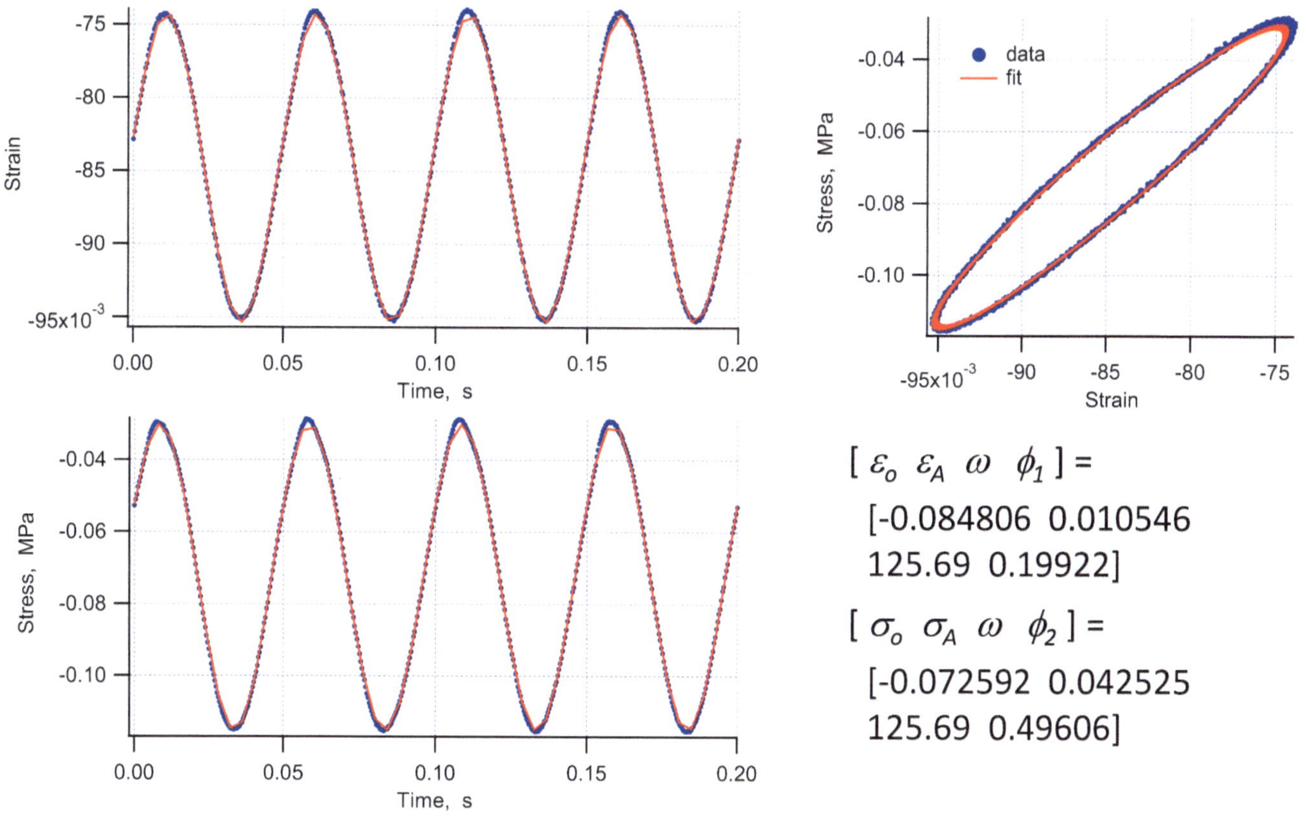

Fig. 16.11 Viscoelastic data analysis of test New05, 0°C, 20 Hz

Take the 20 Hz data of Specimen New05 that was tested at 0 °C for example, the analysis is shown in Fig. 16.11. The sinusoidal functions that best fit the strain and stress data are:

$$\begin{aligned} \varepsilon &= -0.084806 + 0.010546 \ \sin(125.69t + 0.19922) \\ \sigma &= -0.072592 + 0.042525 \ \sin(125.69t + 0.49606) \ \text{MPa} \end{aligned} \tag{16.3}$$

Table 16.2 Viscoelastic parameters calculated from experimental data

Specimen	Temperature, C	Frequency, Hz	δ, rad	*El*, MPa	*E2*, MPa
New04	20	20	0.15	2.06	0.32
	20	40	0.16	2.20	0.36
	20	60	0.22	2.26	0.50
New05	20	1	0.10	1.70	0.17
	20	20	0.16	2.00	0.33
	20	40	0.19	2.08	0.41
	20	60	0.24	2.16	0.52
	0	20	0.30	3.86	1.18
	0	60	0.42	4.16	1.83
	−20	20	0.45	8.90	4.25
	−20	40	0.49	10.48	5.55
	−20	40	0.49	10.29	5.51
	−20	60	0.55	10.93	6.69
	−40	20	0.27	91.96	25.65
	−40	20	0.27	94.89	26.00
	−40	40	0.24	107.09	25.98
	−40	40	0.18	104.85	18.95
	−40	60	0.30	104.65	32.02
	−40	60	0.44	98.13	46.35
	−40	60	0.16	106.35	17.31
	−40	60	0.24	102.58	25.58
Aged05	20	20	0.13	1.68	0.22
	20	40	0.16	1.77	0.29
	20	60	0.19	1.78	0.35
	0	20	0.19	2.10	0.40
	0	40	0.30	2.58	0.79
	0	60	0.32	2.67	0.87
	−20	20	0.39	4.44	1.81
	−20	40	0.46	5.27	2.60
	−20	60	0.51	5.70	3.17
Aged06	20	20	0.14	1.71	0.24
	20	20	0.14	1.75	0.25
	20	40	0.17	1.80	0.31
	20	40	0.17	1.76	0.30
	20	60	0.22	1.83	0.42
	0	20	0.24	2.46	0.59
	0	20	0.23	2.48	0.58
	0	40	0.31	2.81	0.88
	0	60	0.39	2.93	1.20
	−20	20	0.45	6.05	2.90
	−20	40	0.49	6.83	3.68
	−20	60	0.63	7.61	5.50

The strain phase lag is

$$\delta = 0.49606 - 0.19922 = 0.29684 \text{ rad} \tag{16.4}$$

The storage and loss modulus are

$$\begin{aligned} \text{E1} &= (0.042525/0.010546)\cos\ (0.29684) = 3.856 \text{ MPa} \\ \text{E2} &= (0.042525/0.010546)\sin\ (0.29684) = 1.1795 \text{ MPa} \end{aligned} \tag{16.5}$$

The calculated values of δ, $E1$, and $E2$ are listed in Table 16.2. The effects of temperature and frequency on these parameters are shown in Fig. 16.12. For the temperatures tested, the largest phase lag occurs at −20 °C; both storage and loss moduli

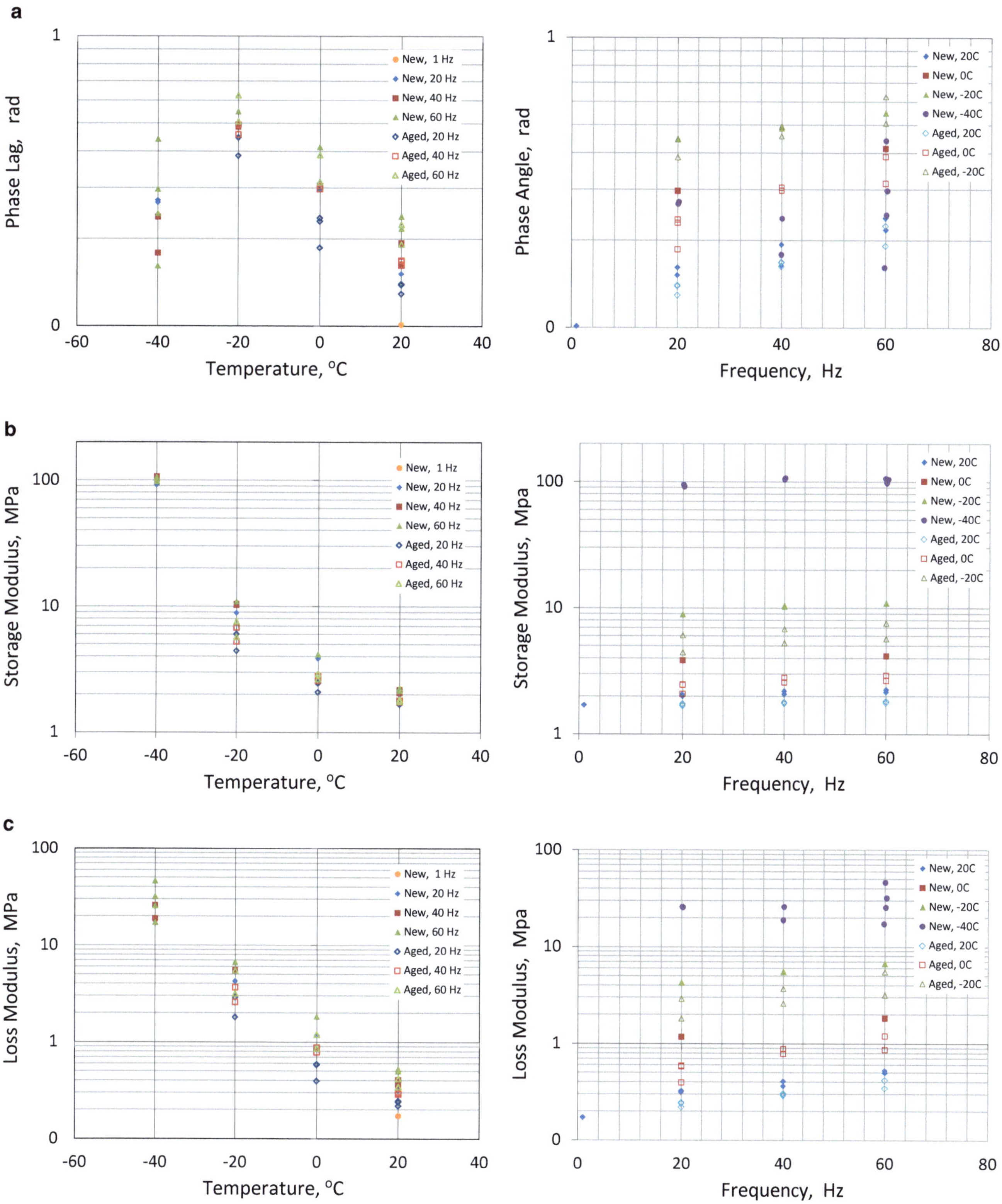

Fig. 16.12 Temperature and frequency effects on (**a**) phase angle, (**b**) storage modulus, and (**c**) loss modulus

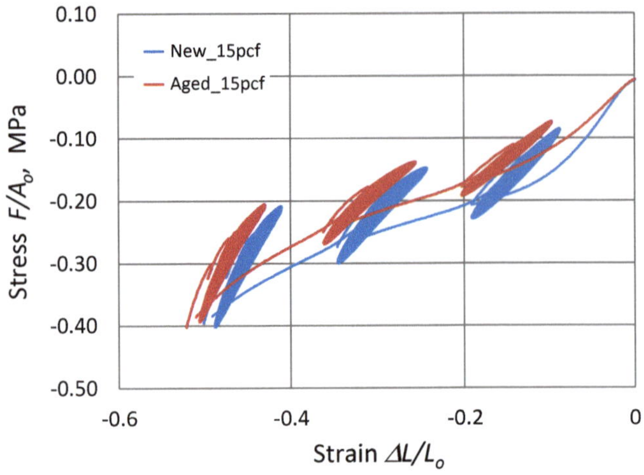

Fig. 16.13 Stress-strain curves and stable hysteresis loops of new and aged 15 pcf foams

increase rapidly as the temperature decreases. All these parameters show increase in value as the frequency become higher. Comparing the new and the aged specimens, in general, the values are lower for the aged foam.

16.7 Pre-Strain Beyond the Linear Viscoelastic Regime

Considering 20 Hz, 1.0 % strain amplitude loading profile, and various pre-strains beyond the linear elastic regime, initial dynamic characterization of the new and aged foams was done at room temperature. Three pre-strains for new foam specimen test were −13.9, −29.5, and −45.6 %; three for aged foam were −15.0, 30.9, and 47.5 %. The stress strain behaviors are plotted in Fig. 16.13. More investigation is needed.

16.8 Conclusion

Dynamic characterization of the initial linear viscoelastic regime of the new and aged 15 pcf flexible polyurethane foams were performed. The effects of temperature and loading frequency on storage and loss moduli are obtained. Data show that foams become stiffer and dissipate more energy at lower temperature and higher frequency. Since the aged and new foam specimens were from different batches and many other factors could influence the material property, e.g. density, cell structure orientation, etc., the observed differences between these two foams should not be interpreted as an aging effect.

The uncertainties increase at low temperatures (<−40 °C) and high frequencies (>40 Hz) for the current setup due to the low load capacity and temperature stability.

Preliminary cyclic experiments of specimens with large pre-strain were studied. More investigation is underway.

Acknowledgement Sandia National Laboratories is a multi-program laboratory managed and operated by Sandia Corporation, a wholly owned subsidiary of Lockheed Martin Corporation, for the U.S. Department of Energy's National Nuclear Security Administration under contract DE-AC04-94AL85000.

References

1. White, S.W., et al.: Experimental techniques and identification of nonlinear and viscoelastic properties of flexible polyurethane foam. Nonlinear Dyn. **22**, 281–313 (2000)
2. generalplastics.com
3. Gibson, L.J., Ashby, M.F.: Cellular Solids: Structure & Properties. Pergamon Press, Oxford (1988)
4. Werner, B.T., Lu, W.-Y., Connelly, K.: Compression Testing of Aged Low Density Flexible Polyurethane Foam. #373. SEM, Bethel (2016)

Chapter 17
Compression Testing of Aged Low Density Flexible Polyurethane Foam

Brian T. Werner, Wei-Yang Lu, and Kevin Connelly

Abstract Flexible open celled foams are commonly used for energy absorption in packaging. Over time polymers can suffer from aging by becoming stiffer and more brittle. This change in stiffness can affect the foam's performance in a low velocity impact event. In this study, the compressive properties of new open-cell flexible polyurethane foam were compared to those obtained from aged open-cell polyurethane foam that had been in service for approximately 25 years. The foams tested had densities of 10 and 15 pcf. These low density foams provided a significant challenge to machine cylindrical compression specimens that were 1 "in height and 1" in diameter. Details of the machining process are discussed. The compressive properties obtained for both aged and new foams included testing at various strain rates (0.05. 0.10, 5 s^{-1}) and temperatures (−54, RT, 74 °C). Results show that aging of flexible polyurethane foam does not have much of an effect on its compressive properties.

Keywords Polymeric foam • Aging • Temperature dependence • Strain rate dependence • Flexible foam

17.1 Introduction

Flexible open cell foams are used widely in packaging to alleviate the effects of shock and vibration during shipping and transportation. As these materials age the cell walls may be exposed to ultraviolet light and the cell walls may oxidize which can lead to stiffening and a reduction in performance. In this study, two low density flexible polyurethane foams were pulled out of service after approximately 30 years and tested against fresh equivalent foam to determine the effects, if any, of aging. The fresh foam is not meant to be a direct comparison as it is not from the same batch and may have slightly different properties. The results are compared on a more broad quantitative and qualitative scale to see whether there are significant changes in material behavior over time. For low density flexible foams, machining adequate specimens to produce quality experimental results is not trivial due to their low stiffness. Manufacturing attempts and solutions are presented that may assist further study of similar soft materials. Since the foams are typically used in shipping and transportation, they are tested across a wide temperature range (−54, 25, 74 °C) in both the rubbery and glassy regimes ($T_g \approx 0$ °C) across three decades of strain rate (0.05, 0.10, 5 s^{-1}). Two different densities of foam were tested, 10 and 15 pounds per cubic foot (pcf).

17.2 Specimen Preparation

One of the main challenges faced in this study was the machining of the aged and fresh foam. Since the density and stiffness of the foam was very low, any force from the cutting tool would simply deform the material instead of cutting it. It was also a challenge to hold the foam in such a way that it would not be compressed significantly prior to testing. Techniques such as cooling the specimen to improve its stiffness or using a hot wire cutting method were considered but not chosen out of concern for influencing the material's measured properties. Attempts at removing specimens using a vibrational hole cutter were unsuccessful due to the applied vertical force deforming the foam. The quantity of aged foam was limited so using a CNC mill to remove a circular path of material around the specimen would have led to significant waste. Conventional techniques used on the lathe were likely to grab the foam and tear it rather than cut it.

B.T. Werner (✉) • W.-Y. Lu • K. Connelly
Sandia National Laboratories, Livermore, CA 94551-0969, USA
e-mail: btwerne@sandia.gov

Y. Zhu, A.T. Zehnder (eds.), *Experimental and Applied Mechanics, Volume 4*,
Conference Proceedings of the Society for Experimental Mechanics Series, DOI 10.1007/978-3-319-42028-8_17

The eventual machining technique that did produce positive results involved a lathe and a high speed (30 k RPM spindle). The foam was cut into cubes on a bandsaw close to the final diameter. The specimens were then turned on the lathe with two live centers. The specimen was held in place using sandpaper adhered to the live center and a small amount of compression. Then the high speed spindle was brought in to abrasively remove the excess material. Once the final diameter was cut, a bandsaw was used to cut the loading faces so they were parallel to one another and square to the loading axis. The final specimen dimensions were nominally 1 in. (25.4 mm) in diameter and 1 in. (25.4 mm) in length.

17.3 Experimental Procedure

Compression testing was performed on both the aged and fresh foam at three different strain rates (0.05, 0.10, and 5 s^{-1}) and at three different temperatures (−54, 25, and 74 °C). While there is not much of a difference between strain rates of 0.05 and 0.10 s^{-1}, the data was to be compared to experiments that were run when the aged foam was fresh. The testing was conducted on a servo-hydraulic MTS Bionix load frame that was equipped with an environmental chamber produced by Applied Testing Systems. The steel loading platens were threaded on to steel push rods. The push rods were kept as short as possible to allow them to extend into the chamber and keep their rigidity at high loads. The specimen strain was measured by using a laser extensometer. The reflective targets were placed on the edge of each of the loading platens. This strain measurement averages over the entire length of the specimen and neglects any strain localization. Due to the study's comparative nature between the fresh and aged foam, consistency in measurement was prioritized. Each specimen was compressed under displacement control to 70 % strain. A picture showing the experimental setup can be seen in Fig. 17.1.

For both the elevated and sub-ambient tests, it is important that the temperature gradient is minimized within the specimen. To determine the soak time, a type K thermocouple probe was used to measure the air temperature in the chamber. Once it was at the desired temperature, the door was opened and the thermocouple probe was pushed into the center of the specimen. For both the 10 and 15 pcf foams, the required soak time was on the order of 20 min. So for all of the testing performed, a room temperature specimen was placed within the environmental chamber for no less than 20 min before the test was performed. In Fig. 17.2 the initial chamber temperature can be seen at $t = 0$, then the temperature reading dips to around room temperature as it reads the internal specimen temperature, and finally it asymptotically approaches the air temperature of the chamber. A similar test was performed at the sub-ambient temperature.

For both the low strain rates (0.05 and 0.10 s^{-1}), the laser extensometer was used to record the strain in the specimen. However, at the high strain rate (5 s^{-1}) the acquisition rate of the laser was not fast enough, which resulted in an oscillatory displacement measurement as seen in Fig. 17.3. For these tests, the actuator displacement, or stroke measurement had to be used to calculate the strain.

In order to correct for machine compliance, a compliance test was run to determine how to correct the stroke measurement. A relatively rigid piece of steel was placed between the loading platens and a compression test under load control was run to measure the error between the laser extensometer reading and the stroke reading (Fig. 17.4). A power law fit was used to approximate the stroke error.

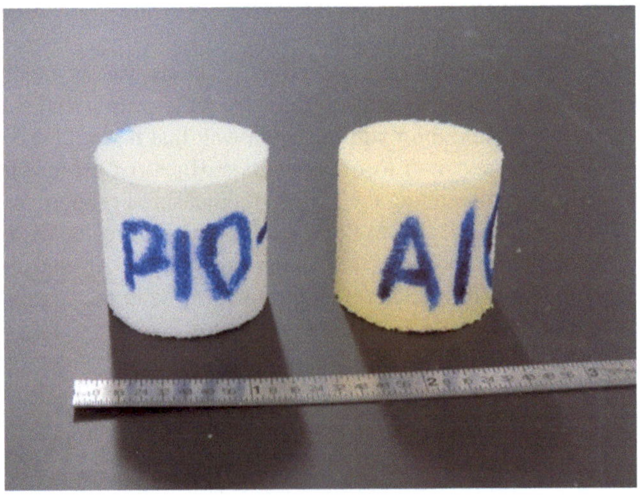

Fig. 17.1 Ten pounds per cubic foot specimens fresh (*left*) and aged (*right*)

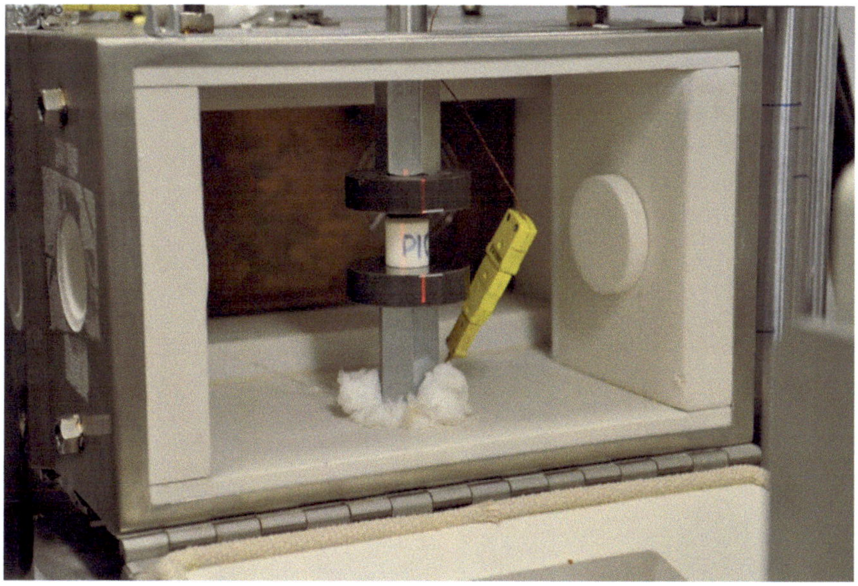

Fig. 17.2 Experimental setup inside the environmental chamber

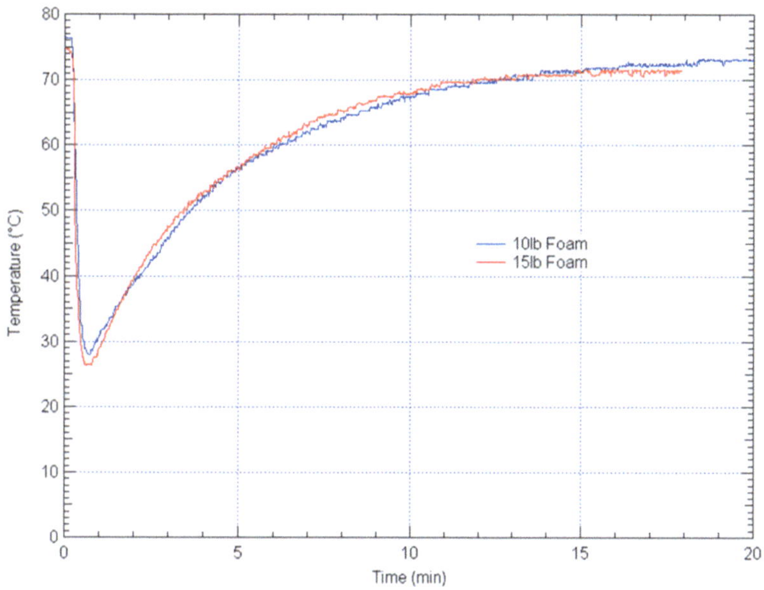

Fig. 17.3 Time for specimen temperature equilibration

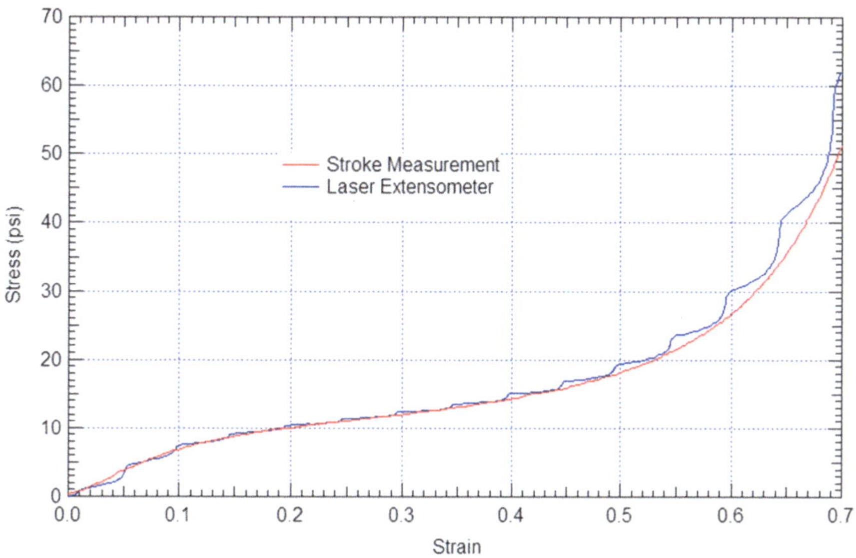

Fig. 17.4 Oscillatory strain measurement from laser extensometer

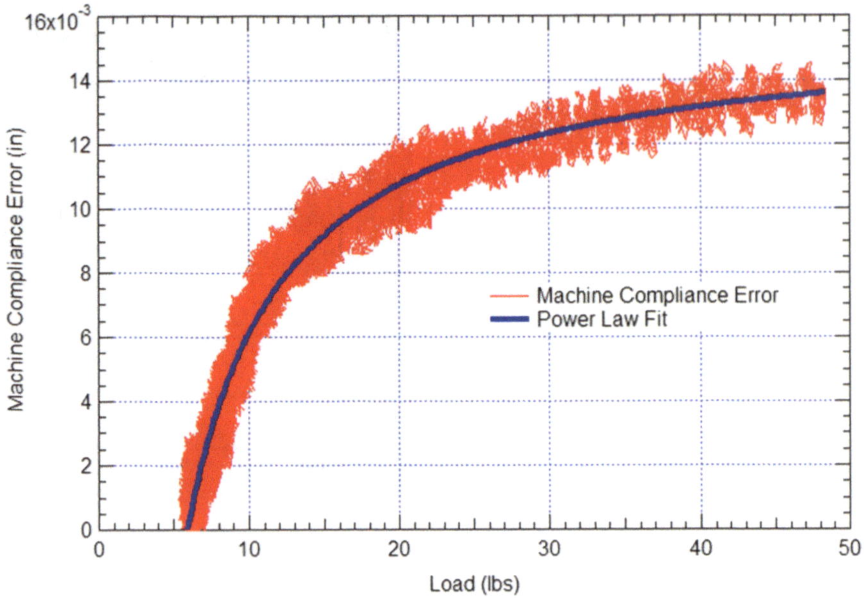

Fig. 17.5 Difference between stroke and laser extensometer reading vs. applied load

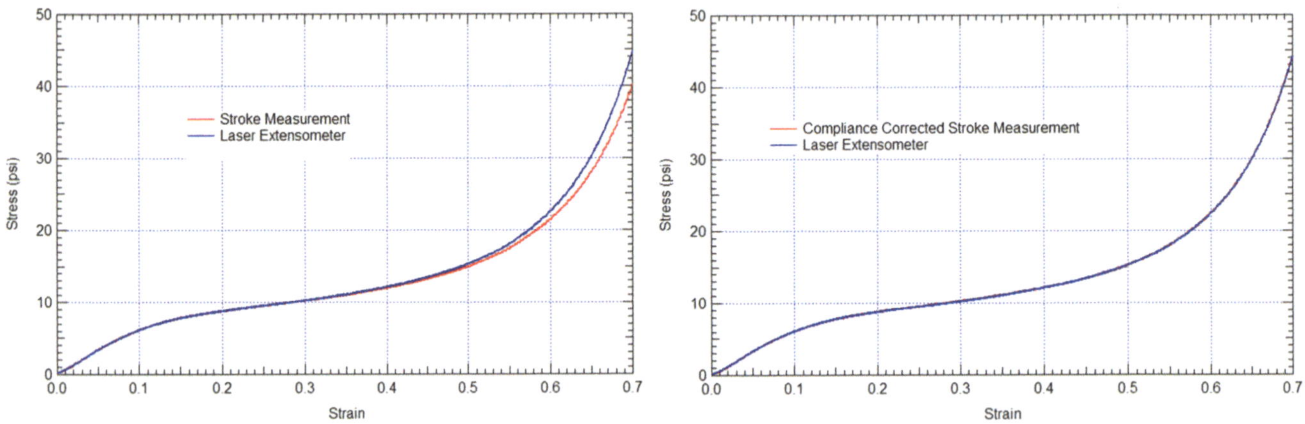

Fig. 17.6 Compliance error between stroke measurement and laser extensometer measurement (*left*) and compliance corrected stroke measurement and laser extensometer (*right*)

Comparing the results from two lower strain rate tests shows how correcting the stroke measurement can provide similarly accurate results as the laser extensometer (Fig. 17.5).

The compliance corrected stroke measurement is equivalent to the laser extensometer reading on the lower strain rate test so when it is applied to the higher strain rate test, shown in Fig. 17.3, the results are similarly accurate (Fig. 17.6). The corrected strain measurement passes right through the center of the oscillatory strain measurement given by the laser extensometer.

Also of concern for the higher strain rate experiment is the acceleration of the actuator. Since it is moving at a velocity of 5 in/s a gap must be given to allow the platen to achieve that speed before it makes contact with the specimen. Tests were conducted to determine over what distance this acceleration occurred and care was taken to ensure that this gap was left at a minimum between the specimen and loading platen before the test was started. Also of concern was the deceleration distance. At high strains the load increases exponentially and it was critical that the two platens did not collide. In order to maintain a constant strain rate of 5 s^{-1} over the strain of interest, the actuator was commanded to over compress the specimen to approximately 80 % strain so that the deceleration of the actuator did not impact the data. This additional compression data was neglected (Fig. 17.7).

Fig. 17.7 Compliance corrected stroke measurement for high strain rate test

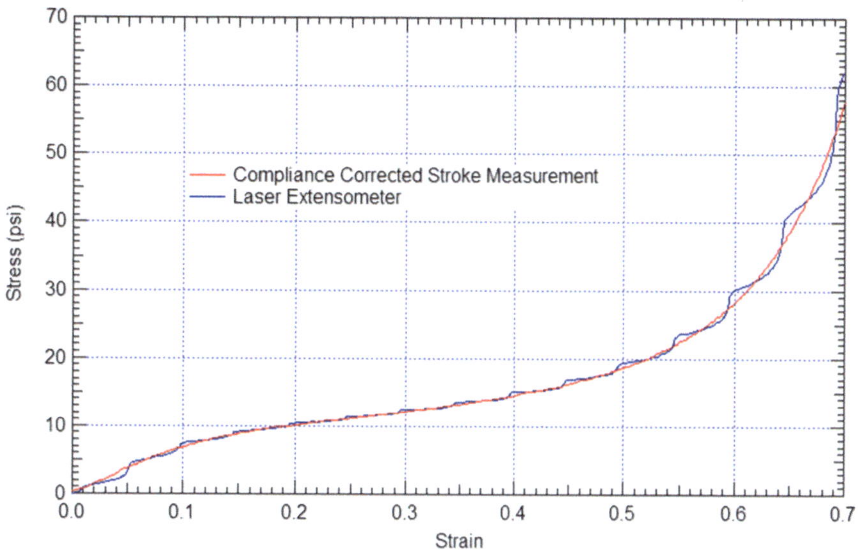

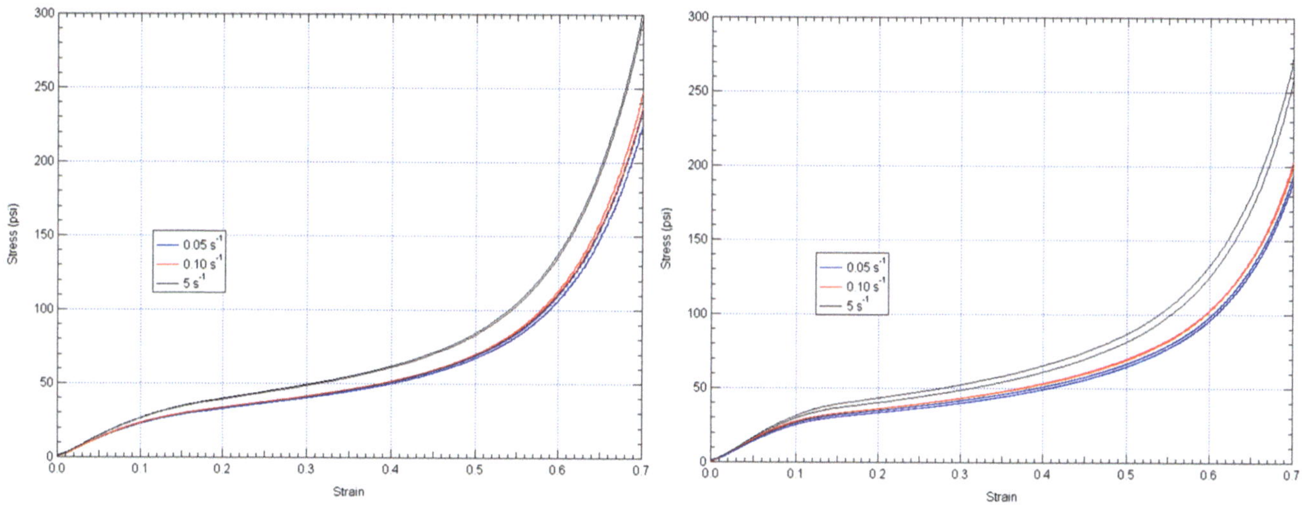

Fig. 17.8 Compression results at 25 °C for 15 pcf foam aged (*left*) and fresh (*right*)

17.4 Results

Qualitatively, the fresh and aged foams behaved very similarly. At each of the temperatures and at each of the strain rates, there was no significant difference between the responses. Stress-strain curves for the 15 pcf foam are shown in Figs. 17.8, 17.9 and 17.10 at each of the three temperatures tested. While there are some differences in some of the stresses, typically for a given strain the stress in the aged foam is within 10–15 % of that of the fresh foam. These differences cannot solely be attributed to aging either due to both batch variation and scatter from specimen to specimen. While only the 15 pcf stress-strain curves are shown here for brevity, the 10 pcf stress-strain curves were also similar between the aged and fresh foam.

A good comparison to illustrate the difference between the fresh and aged foam can be found by looking at the moduli recorded for each temperature and strain rate in Tables 17.1 and 17.2. While the main concern was that the aged foam would show stiffening behavior over time, it consistently showed a lower modulus than the fresh foam. Even if the fresh foam is not a good analog for the aged foam when it was new, there are no significant differences in how both aged and fresh foams dissipate energy. DMA tests performed at Sandia National Laboratories also show that the fresh foam is stiffer than the recovered aged foam [1].

The trends seen between strain rate and temperature appear to be consistent for both the fresh and aged foams. There does not appear to be a detrimental effect on the foam due to aging.

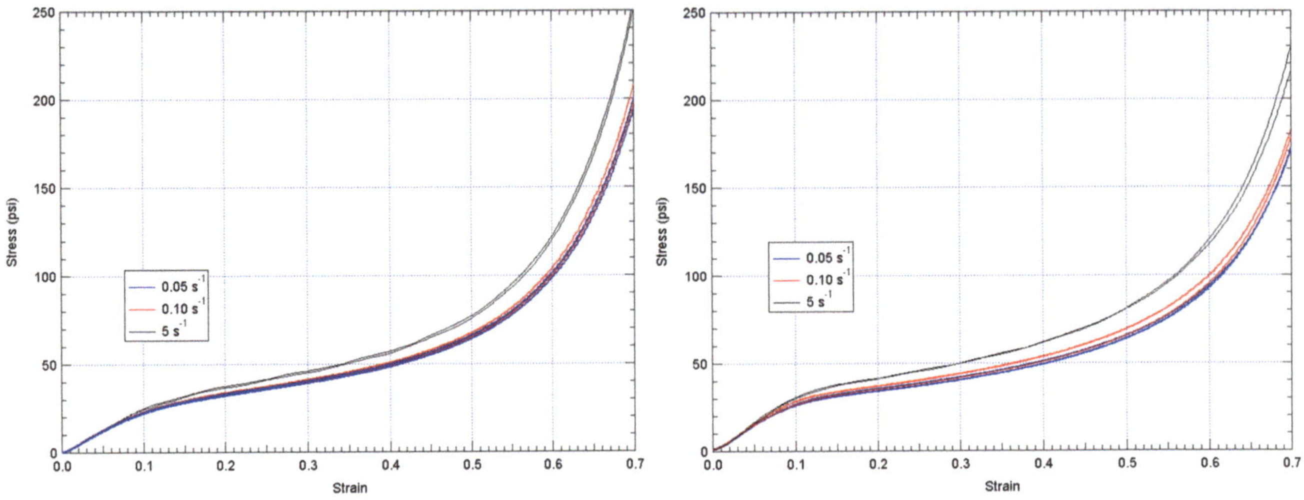

Fig. 17.9 Compression results at 74 °C for 15 pcf foam aged (*left*) and fresh (*right*)

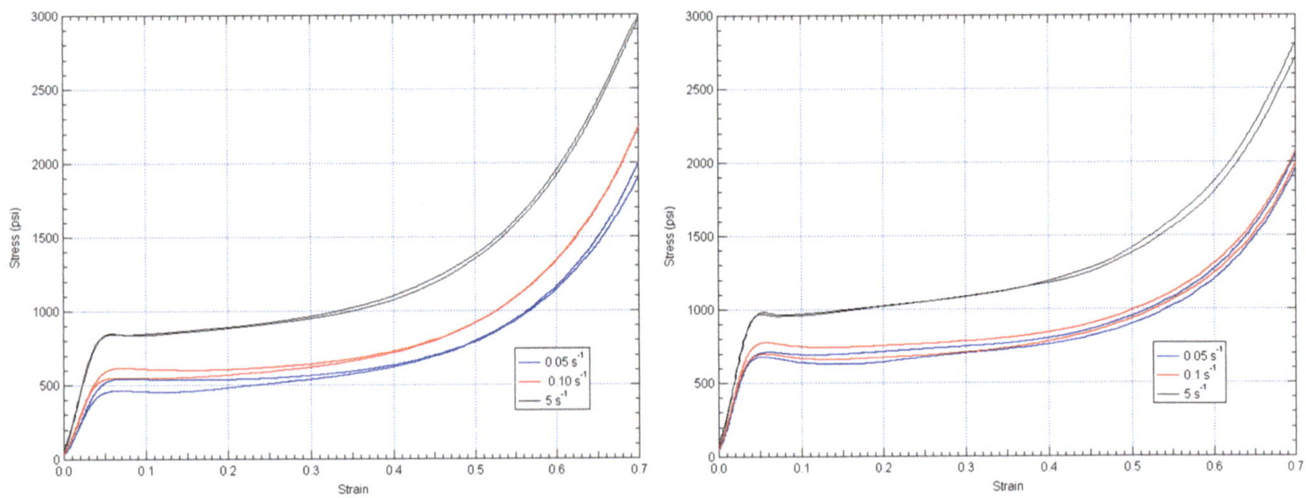

Fig. 17.10 Compression results at −54 °C for 15 pcf foam aged (*left*) and fresh (*right*)

Table 17.1 Modulus for 10 pcf foam

	Aged			Fresh		
Modulus (psi)	Strain rate (s^{-1})			Strain rate (s^{-1})		
Temperature (°C)	0.05 s^{-1}	0.10 s^{-1}	5 s^{-1}	0.05 s^{-1}	0.10 s^{-1}	5 s^{-1}
−54	5260	5760	8150	6890	6670	8940
25	61	64	72	67	70	79
74	61	66	75	57	60	63

Table 17.2 Modulus for 15 pcf foam

	Aged			Fresh		
Modulus (psi)	Strain rate (s^{-1})			Strain rate (s^{-1})		
Temperature (°C)	0.05 s^{-1}	0.10 s^{-1}	5 s^{-1}	0.05 s^{-1}	0.10 s^{-1}	5 s^{-1}
−54	12,200	14,200	22,100	19,500	20,800	24,200
25	267	273	306	322	334	360
74	263	266	287	319	325	355

17.5 Summary

The motivation behind this study was to determine whether the aged foam's properties have changed significantly over the 30 years that it has been in service. It was compared with a similar polyurethane foam material that was produced recently. Initial results show that there has been no significant aging while it has been in use. Significant oxidation or damage from ultraviolet light would most likely make the foam stiffer and more brittle. However, the aged foam was actually not as stiff as the new foam. There are a few possible reasons for this. The most significant is that these are not the same exact foams. It is unknown whether the processing techniques have changed since the original foams were made. A true comparison would be between foams from the same batch. Another issue is that it is unknown whether these foams display any anisotropy. The aged foams have no directionality associated with them so it would be hard to compare. It is also possible that the aging was limited to the surface of the aged foam. Machining may have removed the stiff, brittle, aged outer layer and exposed a fresher internal material. Qualitatively, there does appear to be more discoloration on the surface of the aged foam. When the material is used as packing material though, it is the bulk that determines how well it can absorb shocks and the internal material performed quite well. From the results of this study, the aging of the foam may not be as critical as some may fear. Further study to determine whether there is an aging gradient towards the surface may be needed.

Acknowledgments The work of April Nissen and Jose Morfa at Sandia National Laboratories helped to show that while the results may be counterintuitive, they are consistent with other measurement techniques on this particular material.

Sandia National Laboratories is a multi-program laboratory managed and operated by Sandia Corporation, a wholly owned subsidiary of Lockheed Martin Corporation, for the U.S. Department of Energy's National Nuclear Security Administration under contract DE-AC04-94AL85000.

Reference

1. Morfa, J. Jr., Nissen, A., Werner, B.T.: DMA and Compression Tests on Flexible Polyurethane Foam. Sandia National Laboratories, Livermore (2015)

Chapter 18
Viscoelastic Relaxation of HEMA-DMAEMA Responsive Hydrogels

Chandler C. Benjamin, Roderic S. Lakes, and Wendy C. Crone

Abstract Responsive hydrogels are a class of shape memory materials that undergo a large elastic volumetric change when interacting with a stimulus and can return to their original shape when that stimulus is removed. Important to their use in these device applications are the fundamental mechanical properties of these materials. The pH-sensitive 2-hydroxyethyl methacrylate, 2-(dimethylamino) ethyl metha-crylate (HEMA-DMAEMA) hydrogel is a linear-viscoelastic polymer that is used in microfluidics because it can be easily polymerized within these devices and their special stimuli responsive capabilities making them ideal candidates for sensors and actuators. In the experiments described in this paper the stress relaxation due to a step strain is studied. The rise time for our experiments was 0.3 s and the step-strains ranged from 3 to 7 % strain. Relaxation was recorded over three decades of time (1, 10, 100 and 1000 s). It was found that within this range the HEMA-DMAEMA hydrogel displayed linear-viscoelastic behavior.

Keywords Microfluidics • Hydrogel • Viscoelasticity • Interfacial

18.1 Introduction

Stimuli-responsive hydrogels are a class of three-dimensional cross-linked elastomers that exhibit reversible volumetric changes when introduced to an external stimulant, such as pH. This process is realized because the hydrogel is able to convert chemical energy into mechanical energy. Swelling in hydrogels is diffusion controlled and diffusion times are proportional to length squared. Thus when dimensions are on the millimeter to micrometer scale diffusion times are dramatically reduced. This reduction in response time coupled with the reversibility of the volumetric change makes the stimuli-responsive hydrogel an ideal materials for use in microfluidic devices [1].

Microfluidic design and device development is often conducted with limited input on the mechanical capabilities of the materials. In these microfluidic environments wear, fatigue, friction and adhesion can become important design considerations. As the field moves beyond the proof-of-concept devices, more control over the materials design and a fundamental understanding of the materials properties is required. For this reason a basic understanding of the material properties of these hydrogels and their behavior as design components are essential to successful application [2–5].

Here we examine the viscoelastic relaxation of the pH-sensitive 2-hydroxyethyl methacrylate, 2-(dimethylamino) ethyl metha-crylate (HEMA-DMAEMA) hydrogel. The viscoelastic characterization of the (HEMA-DMAEMA) hydrogel was inspired by recent experiments conducted by this lab group seeking to study the adhesive properties of the hydrogel [6].

18.2 Viscoelasticity

When strains are small, most solid bodies can be described using Hooke's law constitutive equation. This gives a linear relationship between stress and strain that is governed by a modulus of elasticity, $\sigma = E\varepsilon$, where E is the Young's modulus of the solid body. The inverse relationship, describing strain in terms of stress, is given by, $\varepsilon = J\sigma$. Where J is the compliance of

C.C. Benjamin
Engineering Mechanics Program, University of Wisconsin–Madison, 1500 Engineering Drive, Madison, WI 53706, USA

Department of Engineering Physics, University of Wisconsin–Madison, 1500 Engineering Dr., Madison, WI 53706, USA

R.S. Lakes • W.C. Crone (✉)
Department of Engineering Physics, University of Wisconsin–Madison, 1500 Engineering Dr., Madison, WI 53706, USA
e-mail: crone@engr.wisc.edu

Y. Zhu, A.T. Zehnder (eds.), *Experimental and Applied Mechanics, Volume 4*,
Conference Proceedings of the Society for Experimental Mechanics Series, DOI 10.1007/978-3-319-42028-8_18

the solid body and $J = \frac{1}{E}$. On the other side of the spectrum a viscous fluid that is subject to a shear stress has the following constitutive relationship, $\sigma = \eta \frac{d\varepsilon}{dt}$, where η is the viscosity of the fluid and the stress depends not on the strain but the strain rate. All real materials, to some degree, are not exactly described as solid or fluid. Most materials exhibit viscous and elastic characteristics. A viscoelastic material is described by having its stress and strain relationship depend on time.

Viscoelastic behavior becomes apparent when the material displays creep, or an increase in strain when the stress is held constant or when the material displays relaxation or a gradual decrease in stress for a given strain that is held constant. These two phenomena may be used as experimental methods used to characterize materials. They are also of concern in the design and application of these materials [7]. Viscoelastic stress relaxation is the method used for characterization of the HEMA-DMAEMA hydrogel in this experimental work.

On the molecular level, the relaxation in polymers is due to molecular rearrangement [8]. The pH-sensitive hydrogel 2-hydroxyethyl methacrylate (HEMA) is a cross-linked elastomer. The elasticity of these hydrogels results from two different mechanisms, changes in conformational entropy and internal energy. For cross-linked elastomers, the energetic contribution is negligible in comparison to the entropic contribution and is typically ignored. After a polymer network is straightened, or compressed, this process is followed by a reconfiguration of the network to achieve maximum conformational entropy.

To determine whether a viscoelastic materials displays linear vs. non-linear behavior the isochronal stress vs. strain plots can be examined. In evaluating such results, a linear slope to the curve indicates that the material is linearly viscoelastic, while a non-linear curve indicates a non-linear viscoelastic behavior.

18.3 Materials and Methods

18.3.1 Sample Fabrication

Using the experimental techniques previously developed for the mechanical characterization of responsive hydrogels [9], pH-sensitive 2-hydroxyethyl methacrylate, 2-(dimethylamino) ethyl metha-crylate (HEMA-DMAEMA) (n=10) hydrogels were polymerized into samples using a polydimethlsiloxane (PDMS) mold (Fig. 18.1).

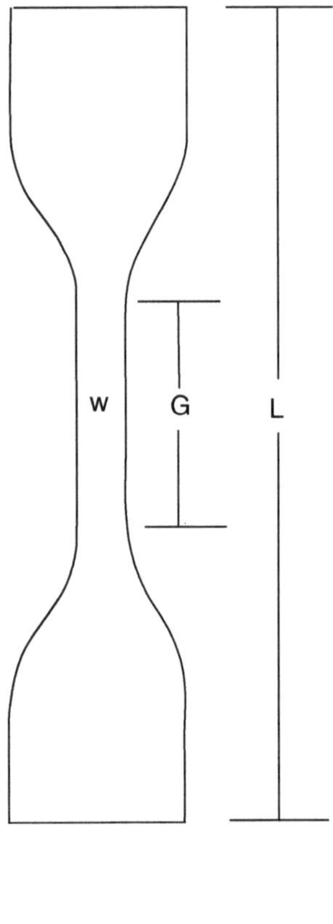

Fig. 18.1 A type IV ASTM D 638-99 tensile testing diagram. W is the gauge width of 1.5 mm, G is the gauge length of 8.0 mm and L is the total sample length of 40 mm. The thickness of each sample is 1.00 mm

A type IV sample size was used based on ASTM D 638-99 standards [10]. Samples were fabricated with a total length of 40 mm, a gauge length of 8.0 mm, a gauge width of 1.5 mm and a thickness of 1.0 mm. A PDMS stencil was produced by a compression molding process that makes use of a negative-tone UV photoresist silicon wafer with the requisite pattern.

The PDMS stencil was placed on top of a flat PDMS substrate. The stencil channels were filled with the HEMA-DMAEMA pre-polymer solution. A third layer of PDMS was placed on top of the stencil. The pre-polymer solution is then exposed to UV light at an intensity of 20 mW/cm^2 for a total of 120 s [9]. After polymerization the samples were removed and washed with methanol and deionized water. The samples were then stored in a phosphate buffered saline solution (PBS) for twenty-four prior to testing.

18.3.2 Mechanical Testing

The fastest rise time that could be achieved was 0.3 s with the equipment available. This prevented accurate capture of relaxation data for any decade of data before the one second mark. The samples were tested on an Instron model 5548 screw driven micro testing apparatus with a 10 N load cell inside a custom built environmental chamber. The displayed position is accurate to with $\pm$ 6 μm over 100 μm of travel. The loads recorded are accurate to within $\pm$ 0.25 % [11] of displayed force. Further details on the construction of the environmental chamber used for tensile tested can be found in Johnson et al. [9].

The HEMA-DMAEMA hydrogel samples must be tested in an aqueous environment for the duration of the test. The hydrogel samples are swollen in phosphate buffered saline (PBS) solution at a pH of 7.0 for 24 h prior to testing. The samples are then tested in PBS for the duration of the test. The pH of 7.0 causes no additional swelling of the pH sensitive hydrogel outside of the normal diffusion of H_2O into the hydrogel. The samples were strained from 3 to 7 % strain at a rate of 1000 mm/min with a rise time of 0.3 s.

Three different samples were tested from different batches of pre-polymer HEMA-DMAEMA solution. The first sample was fabricated too thin (0.5 mm) and the data recorded was on the same order of magnitude as the background fluctuations of the load cell. Two samples used for the analysis below were fabricated with a thickness of (1.0 mm). This thickness resulted in recorded loads that were several orders of magnitude greater than the load tolerance of the 10 N load cell. One test was conducted at each step strain ranging from 3 to 7 % for each sample tested.

18.3.3 Stress Relaxation

Time dependent stress response to a step function in strain is stress relaxation. To mechanically characterize the viscoelastic properties of a material one may measure the relaxation modulus, the creep compliance, and the internal damping of the material known as tan(δ). For linear viscoelastic materials there is an Laplace transform relationship between the relaxation and creep, $J(s)E(s) = 1/s^2$, and only one function is needed to calculate the other. There is also a relationship between the relaxation modulus and the dynamic modulus as well as a relationship between the dynamic modulus and tan(δ). Thus determination of the relaxation or creep, for linear viscoelastic materials, yields all the other quantities of interest. The HEMA-DMAEMA material is linearly viscoelastic and in these experiments the relaxation modulus is determined. The general linear viscoelastic constitutive equation is given by the Boltzman superposition integral,

$$\sigma(t) = \int E[t-\tau]\frac{d\varepsilon(\tau)}{d\tau}d\tau, \tag{18.1}$$

and for a step strain a functional form is given as,

$$\varepsilon = \varepsilon_0 \mathcal{H}(t) \tag{18.2}$$

where $\mathcal{H}(t)$ is the Heaviside function. Using this relationship in Eq. (18.1) we define the relaxation modulus for a step strain as,

$$E(t) = \frac{\sigma(t)}{\varepsilon_o}, \tag{18.3}$$

This gives the relaxation stiffness as a function of time for a given step strain. We note that the relaxation modulus is a function of time and stress, time and step strain $E(t) = f[\sigma, \varepsilon_o; t]$, and a plot of the relaxation modulus against time at various step strains can be used to determine a relaxation modulus.

To model the relaxation we use a two term Prony series with two distinct time constants,

$$E(t) = E_0 + \sum_{i}^{N} E_i exp(-t/\tau_i), \tag{18.4}$$

where E_0 represents the equilibrium modulus in units of pascals, E_i is a fitting constant with units of pascals and τ_i represents the time constant with units of time.

18.4 Results

The relaxation modulus of the HEMA-DMAEMA for 3 to 7 % step strain was determined by calculating the engineering stress from the load vs. time data and then plotting the stress over time. A suitable function is then fit to the stress vs. time data to determine the relaxation modulus. Figure 18.2 gives the measured load vs. time for the step strains tested plotted on a log-log scale. The tests were conducted for 1000 s and the strain rate for the step strain was 1000 mm/min and this gave a rise time of 0.3 s.

In Fig. 18.3 the stress is plotted as a function of strain for the 1, 10, 100 and 1000 s time points. At each time point the plotting of stress vs. strain is linear. Figure 18.3 is representative of one sample tested at 7 % strain.

Figure 18.4 is a representative example the relaxation at 6 % step strain. Two different samples were tested from different batches of HEMA-DMAEMA showed consistent results.

The relaxation modulus is plotted in Fig. 18.5 for 7 % step strain for two different samples.

For the data reported in Table 18.1 two different test samples are represented. Each sample was fabricated from different batches of pre-polymer HEMA-DMAEMA solution. One test one conducted at each step strain from 3 to 7 % on each sample.

18.5 Discussion

This study looked at the viscoelastic relaxation properties of pH-sensitive 2-hydroxyethyl methacrylate, 2-(dimethylamino) ethyl metha-crylate (HEMA-DMAEMA), determining the relaxation modulus as a function of time for a given step strain. Other similar HEMA based polymers have been studied, with the triton methacrylate, 2-hydroxyethyl methacrylate

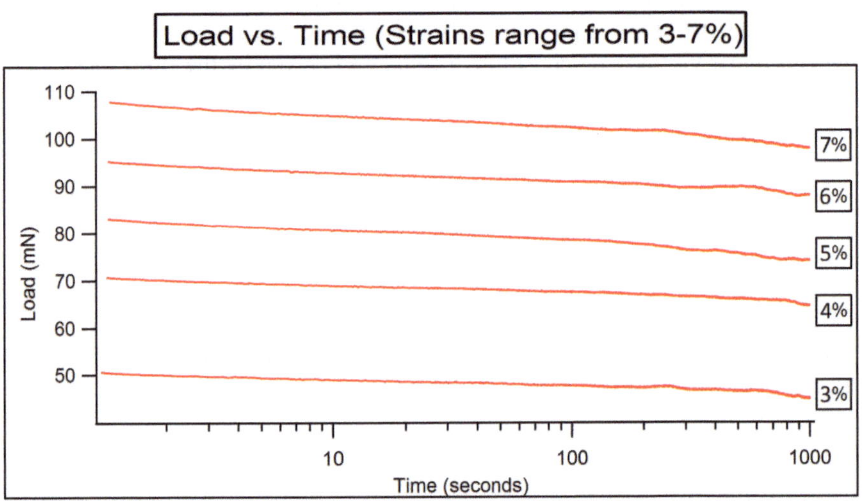

Fig. 18.2 Load vs. Time for the HEMA-DMAEMA responsive hydrogel. Strains range from 3 to 7 %. This graph is representative of one curve for each strain increment. One test was performed at each strain increment

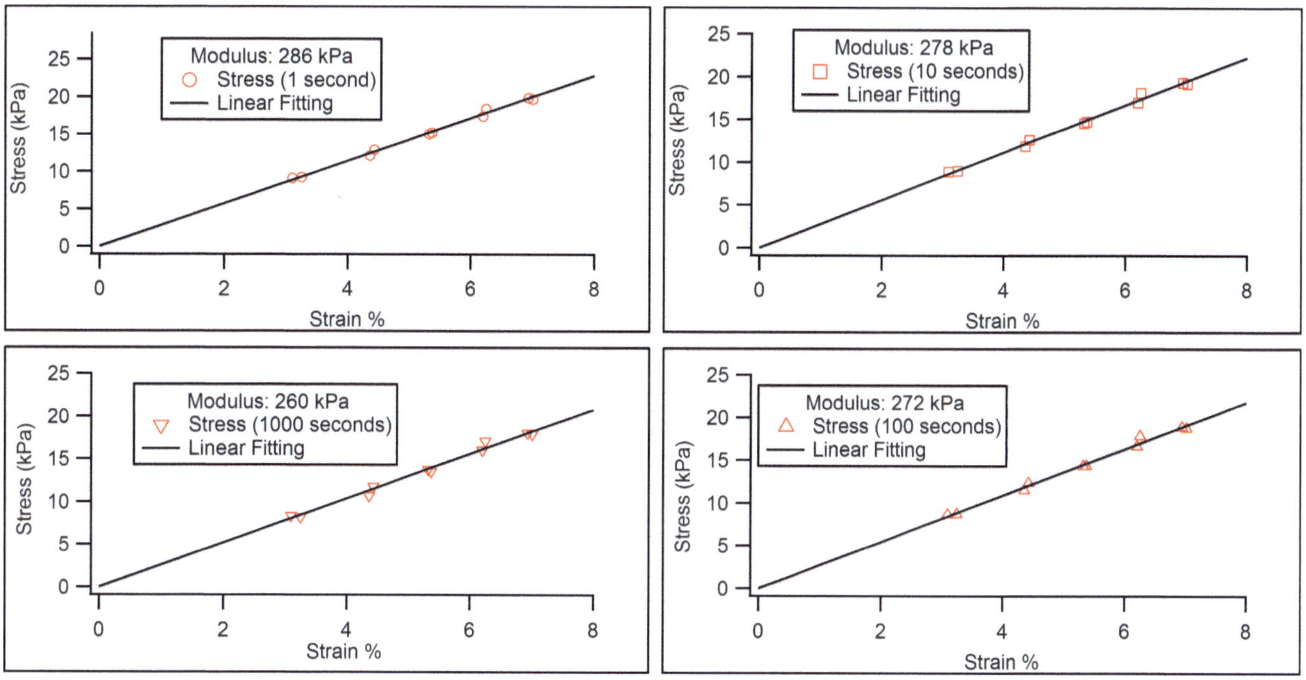

Fig. 18.3 Isochronal stress vs. strain plots for the (1, 10, 100 and 1000) second time points

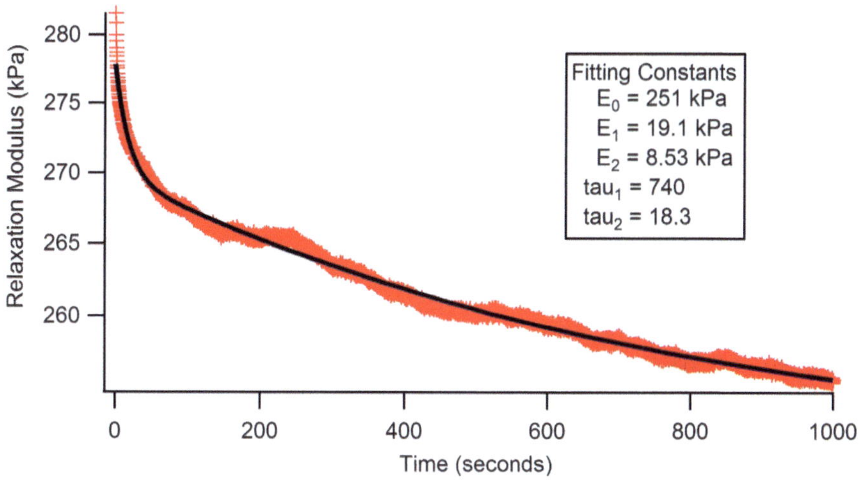

Fig. 18.4 Plot of modulus vs. time. This sample was stretched to 6 % strain and relaxation took place over three decades. E_o

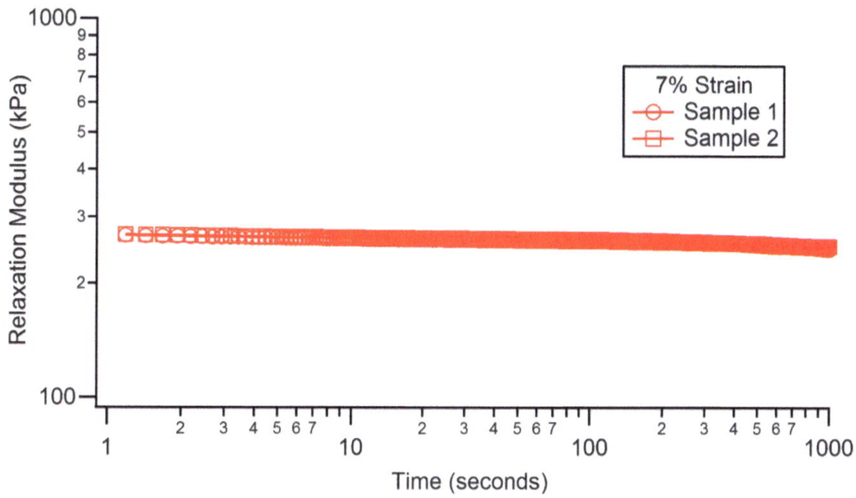

Fig. 18.5 Relaxation modulus plotted for a step strain 7 %

Table 18.1 Relaxation modulus parameters

Samples	E_o(kPa)	E_1 (kPa)	E_2 (kPa)	τ_1	τ_2
1	250 ± 7.51	22.71 ± 6.20	8.58 ± 1.02	1060 ± 442	22.4 ± 8.77
2	249 ± 14.1	27.2 ± 10.1	8.88 ± 1.30	1670 ± 1080	28.1 ± 19.2
Total	249 ± 10.3	25.0 ± 8.25	8.73 ± 1.11	1370 ± 842	25.3 ± 14.4

(MTHEMA) studied by Larraz et al. [13] displaying a power law type relaxation modulus and the poly(2-hydroxyethyl methacrylateco- methyl methacrylate P(HEMA-co-MMA) studied by Brazel and Peppas, [14] displaying and exponential type relaxation modulus.

For the range of step strains tested, from 3 to 7 % strain, the HEMA-DMAEMA responsive hydrogel behaves as a linear viscoelastic material, which is illustrated in Figs. 18.4 and 18.5. The high strain rate required for accurate relaxation tested caused the hydrogel to break when strained over 7 % with the initial step strain. For a linear viscoelastic material the relaxation functions are independent of strain level. A linear viscoelastic material allows characterization of a material with only knowledge of the relaxation or creep functions. In contrast, for a non-linear viscoelastic material the stress relaxation would be a direct function of strain level. There is no indication of a strain hardening or strain softening effect that would be indicative of non-linear behavior in the range of 3 % strain to 7 % strain.

The experimental results demonstrated that the relaxation of the HEMA-DMAEMA hydrogel can be approximated by a two term Prony series with two distinct time constants. Figure 18.4 reports the values describing the relaxation modulus as a function of time and stress, time and step strain. The modulus E_0 represents the modulus of the material after long relaxation times. The magnitude of this modulus (249 ± 10.3 kPa) indicates that the material relaxes for a period of time then plateaus to an equilibrium modulus. This is an indication that the material behaves like a viscoelastic solid. From Fig. 18.4 it can be seen that there is rapid relaxation that occurs over the first decade (1–10 s). This is followed by a longer period of slow relaxation. This is seen in the difference in magnitudes of the two time constants. A smaller time constant indicates rapid relaxation. The time constant τ_2 was calculated as 25.4 ± 14.4 s and represents the rapid initial decay. While the time constant τ_1 was calculated at 1370 ± 842 s and this represented the slow, steady relaxation that occurred after the first decade.

Acknowledgements This research was supported with funds from the Graduate Engineering Research Scholars Program (CCB) and a Vilas Life Cycle Professorship from the University of Wisconsin-Madison, as well as the National Science Foundation Graduate Research Fellowship Program (CCB).

References

1. Beebe, D.J., Mensing, G.A., Walker, G.M.: Physics and applications of microfluidics in biology. Ann. Rev. Biomed. Eng. **4** (1), 261–286 (2002)
2. Oh, K.W., Ahn, C.H.: A review of microvalves. J. Micromech. Microeng. **16** (5), R13 (2006)
3. Gravesen, P., Branebjerg, J., Jensen, O.S.: Microfluidics-a review. J. Micromech. Microeng. **3** (4), 168 (1993)
4. Shoji, S., Esashi, M.: Microflow devices and systems. J. Micromech. Microeng. **4** (4), 157 (1994)
5. Elwenspoek, M., Lammerink, T.S.J., Miyake, R., Fluitman, J.H.J.: Towards integrated microliquid handling systems. J. Micromech. Microeng. **4** (4), 227 (1994)
6. Benjamin, C.C., Springmann, J.C., Chindhy, S.A., Crone, W.C.: Characterization of the interfacial adhesion for responsive hydrogels on substrates. In: Conference Proceeding: American Physical Society (APS), vol. 57 (2012)
7. Lakes, R.S.: Viscoelastic Materials. Cambridge University Press, Cambridge (2009)
8. Ferry, J.D.: Viscoelastic properties of polymers. Wiley, New York (1980)
9. Johnson, B.D., Bauer, J.M., Niedermaier D.J., Crone, W.C., Beebe, D.J.: Experimental techniques for mechanical characterization of hydrogels at the microscale. Exp. Mech. **44** (1), 21–28 (2004)
10. ASTM Standard: D638: Standard test method for tensile properties of plastics. ASTM International, West Conshohocken (2010)
11. Instron Corporation: Instron operator's guide. Technical Report, Instron Corporation (1997)
12. Lakes, R.S.: Viscoelastic measurement techniques. Rev. Sci. instrum. **75** (4), 797–810 (2004)
13. Larraz, E., Elvira, C., Román, J.S.: Design and properties of novel self-curing acrylic formulations for application in intervertebral disks restoration. Biomacromolecules **6** (4), 2058–2066 (2005)
14. Brazel, C.S., Peppas, N.A.: Dimensionless analysis of swelling of hydrophilic glassy polymers with subsequent drug release from relaxing structures. Biomaterials **20** (8), 721–732 (1999)

Chapter 19
Studying the Influence of the Reclaimed Asphalt Pavement (RAP) on Local Deformation Properties of Asphalt Mixtures

M.-C. Teguedi, E. Toussaint, B. Blaysat, M. Grédiac, S. Moreira, and S. Liandrat

Abstract The present study focuses on the experimental investigation of the mechanical response of recycled asphalt mixtures (RAP) using a full-field measurement technique: the grid method. Four hot mixture asphalt (HMA) specimens containing 0, 20, 40, and 100 % of RAP were prepared. Aggregates of different colors were selected in order to distinguish RAP and virgin aggregates. Compression tests were then carried out during which the displacement and strain fields were measured. It was found that the 100 % RAP specimen exhibits highly concentrated strain distributions around the aggregates. Strain is more homogeneously distributed over the specimen surface of virgin mixes. The analysis of the local behavior of RAP showed a local stiffening of the binder along the border of the RAP aggregates. The obtained results highlight the difference in behavior between specimens both at the micro- and the macro-scales.

Keywords Asphalt • Recycled asphalt pavement • Full-field measurement method • Grid method

19.1 Introduction

The recycling of bituminous materials recovered from resurfacing and rehabilitation projects has become a common practice which allows not only to reduce the construction cost of new roads, but also to contribute to the preservation of natural resources. The recycled asphalt materials (RAP) are generally composed of aggregates and aged asphalt binder. The amount of RAP which is commonly used in hot recycled bituminous mixes is regulated by standards and generally lies between 10 and 30 %. Nowadays, a number of technologies are readily available to produce 100 % RAP hot asphalt mixtures [1]. As the asphalt industry move towards the increase of RAP amount in asphalt mixes, questions are raised concerning the effect of RAP use on the mechanical properties of the mixture. The issue that affects directly the performance of HMA containing RAP is related to the blending efficiency between RAP and virgin materials. Indeed, after the mechanical mixing process, some coarse aggregates remain coated to some aged binder and fine materials. They form a high viscosity shell that may acts together with RAP aggregates like a black rock [2]. It is therefore clear that the investigation of the mechanical response of recycled mixture requires the use of suitable measurement tools that provide information at the same scale as the one this phenomenon.

This study aims at investigating the effect of RAP on the mechanical behavior of recycled asphalt mixtures using the grid method. Recent studies have demonstrated the relevance of such a method for characterizing and comparing the mechanical response of asphalt mixtures [3, 4]. For this purpose, cylindrical specimens made of hot mix asphalt mixtures with different RAP contents were prepared. Compression and recovery tests were carried out. The grid method was used to deduce the displacement and strain fields on the front face of the tested specimens. These measurements were used to compare, at different scales, the change in behavior of asphalt as the amount of RAP increases.

M.-C. Teguedi • E. Toussaint (✉) • B. Blaysat • M. Grédiac
Institut Pascal, Clermont Université, Université Blaise Pascal, BP 10448, Clermont-Ferrand 63000, France

CNRS, UMR 6602, Institut Pascal, Aubière, 63171, France
e-mail: Evelyne.TOUSSAINT@univ-bpclermont.fr

S. Moreira • S. Liandrat
Département Laboratoire de Clermont-Ferrand, CEREMA, DterCE, Clermont-Ferrand, France

Y. Zhu, A.T. Zehnder (eds.), *Experimental and Applied Mechanics, Volume 4*,
Conference Proceedings of the Society for Experimental Mechanics Series, DOI 10.1007/978-3-319-42028-8_19

19.2 Materials and Methods

Four Hot Mixtures Asphalt (HMA) incorporating different RAP contents: 0, 20, 40 and 100 % of RAP were considered in this study. The tests were carried on cylindrical specimens supplied by Cerema Direction Territoriale Centre-Est, France; and fabricated according to the Duriez procedure (EN 12697-26, 2012) (cylindrical-shaped specimens with 80 mm diameter). The RAP materials were composed of granite, basalt and gneiss and recovered from the same source. The virgin materials were constituted from limestone aggregates and a virgin bituminous binder (bitumen grade 50/70). The aggregates were selected by considering the difference in their colors: black for recycled aggregates and white for virgin ones (Fig. 19.1a). Hence, after the mixing process, the recycled material could be easily identified, and subsequently their mechanical response could be observed separately during the test (Fig. 19.1b).

The flat front face under study was obtained by cutting the cylindrical specimen parallel to its axis (Figs. 19.1b and 19.2). The dimension of the surface under investigation for all specimens was about 6.5 cm × 7 cm. The tests performed in this study were compression and recovery tests. They were performed in a temperature controlled room at 21 °C, with a displacement rate of 0.01 mm/s. Since the opposite faces of the specimens are not perfectly parallel after the cutting process, a thin rubber sheet was placed between the specimen sides and the compression plates in order to ensure a uniform displacement on the upper face of the specimen. The displacement/strain measurement technique was the grid method. Its principle is to analyze the deformation of a grid. This grid is transferred onto the surface of the specimen beforehand [5, 6]. Thus, the grid is assumed to perfectly follow the deformation of the specimen surface. It is worth noting that the procedure of grid transferring described in [6] involves a curing phase of 40 h at 37 °C. To prevent heat deterioration of the sample, the glue was allowed to cure at room temperature (20 °C) during a polymerization time of 1 week. The next step consisted in acquiring images of this grid by a camera at different stages of the load. These images were finally processed after testing by using an in-house programme. This programmes uses a specific grid image processing procedure which relies on the calculation of the phase change of this regular pattern between the reference and the current stages [5]. Typically, a resolution in strain of some hundreds of microstrains with a spatial resolution equal to 30 pixels are achieved [7]. The camera used here to capture the grid images was a PCO 2000 camera featuring a 14-bit/2048 × 2048 pixel CCD sensor (CCD = charge-coupled device). The number of pixels used to digitize one grid period was equal to 5 pixels.

19.3 Experimental Results and Discussion

19.3.1 Displacement and Strain Fields

In the following, only results obtained on one specimen per type of asphalt are presented: 100 % RAP, 40 % RAP and virgin mix. The strain fields obtained with the grid method at the end of the loading phase are shown in Fig. 19.3. For each specimen, ε_{xx}, ε_{yy} and ε_{xy} are presented. The ε_{xx} strain maps emphasize the global expansion of the specimen along this

Fig. 19.1 (**a**) RAP materials (*black*) and limestone aggregates (*white*) and (**b**) shape of tested specimens

Fig. 19.2 Front face of samples under study

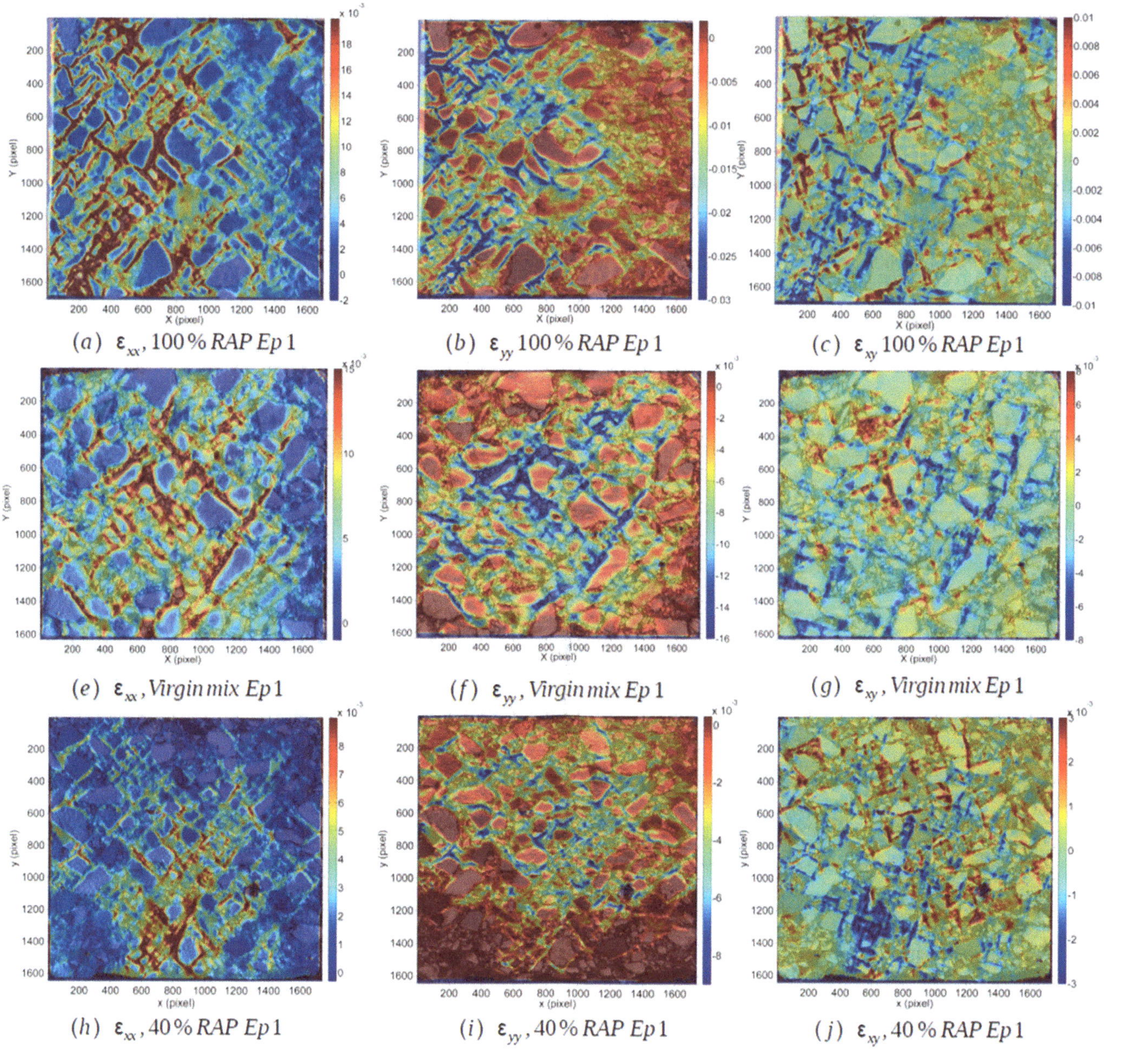

Fig. 19.3 Stain field at maximum load superimposed on the microstructure of samples, for each specimen, ε_{xx}, ε_{yy} and ε_{xy} are shown

direction while a compressive deformation ε_{yy} appears along the vertical direction. It can be observed that the distribution of the coarse aggregates impacts the strain distribution. Indeed, strain is quasi-null in zones corresponding to the aggregates, meaning thus that the whole strain is sustained by the mastic. This was expected due to the significant difference in rigidity between bitumen matrix and aggregates. However, there is no contrast in strain between some small aggregates and mastic. This is due to the fact that their sizes are smaller than the spatial resolution of the measurement. This information means that aggregates displaying a lower size cannot really be distinguished here, and are considered as belonging to the binder.

It can also be observed in Fig. 19.3 that the mix with 100 % RAP shows the highest strain level. The latter is mainly distributed at the bottom part of the specimen. Some localized strain concentrations are observed at the aggregates/binder interface. They can possibly reflect an interface failure in these zones. One the other hand, for the virgin mix, the strain is mainly concentrated in the central part of the specimen and some strain peaks can be observed within the mastic.

19.3.2 Local Behaviour of RAP Aggregates Within the Mixture 40 % RAP

As mentioned in the introduction, the mechanical mixing process does not allow the total liberation of the aged binder. Thus fine particles are still bonded to some coarse RAP aggregates. This section aims at evaluating the effect of the aged asphalt layer on the local strain distribution around the RAP aggregates. The 40 % RAP specimen is therefore studied in this section.

The difference in color between virgin aggregates (white) and recycled ones (black) allows to easily recognize the RAP-aggregates distribution within the area under the study.

Eight regions corresponding to RAP aggregates were selected within the 60 % RAP specimen. The equivalent 2D von Mises strain corresponding the 60 % RAP specimens is considered to observe the strain distribution around these aggregates. Close-up views of the corresponding regions are showed in Fig. 19.4.

The boundary of the recycled aggregates is highlighted in white and the virgin one in blue.

A unexpected phenomenon is that a localized stiffening of the binder occurs around the aggregates in Regions R2, R4 and R8. This phenomenon is particularly pronounced in Region R2 where the lump of RAP surrounding the aggregates remains

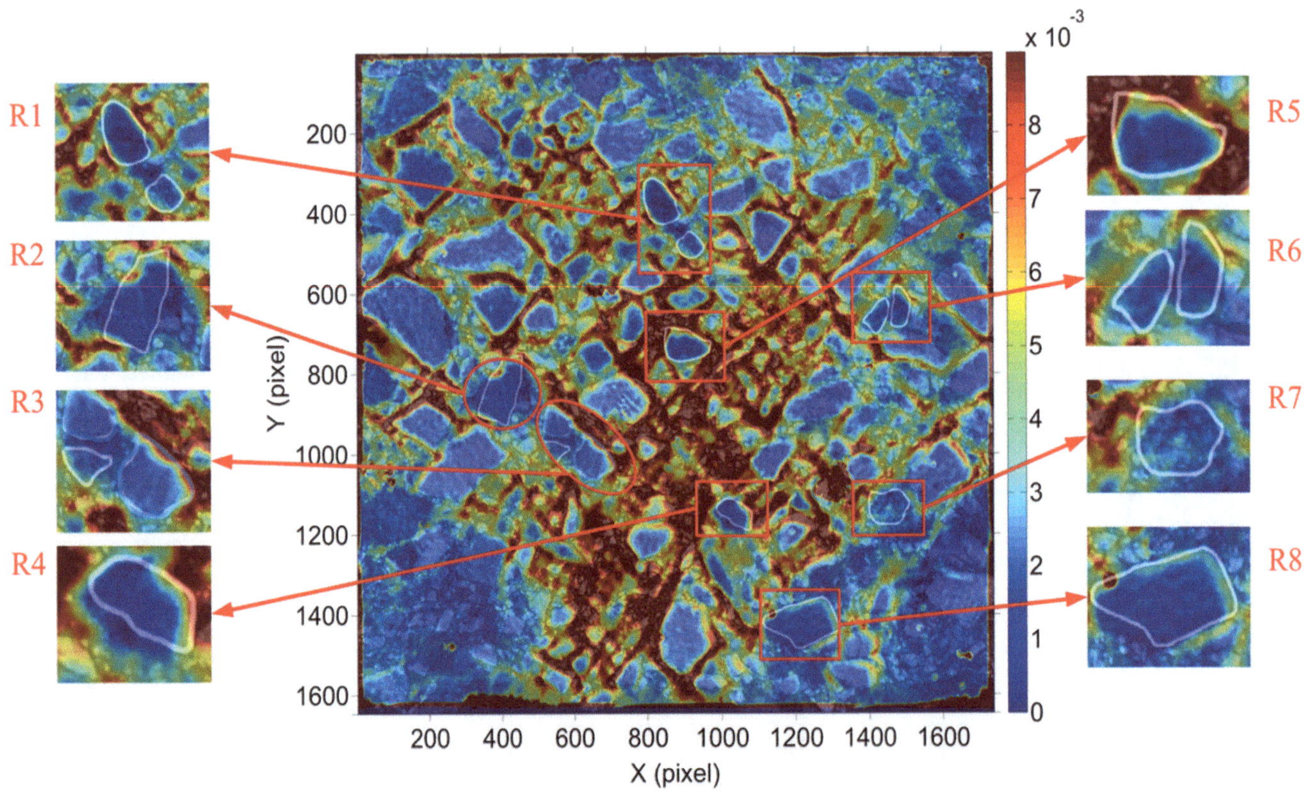

Fig. 19.4 Equivalent von Mises strain distribution within the surface under investigation of the 40 % RAP specimen. Regions R1–R8 are selected, the boundary of the aggregates is highlighted in *white*

practically rigid during the loading phase, and behaves as a part of these aggregates. Similar remarks can be drawn for the RAP lumps located in Region R8. These set of particles remains underfomed during loading phase and follow a rigid body motion. Thus their movement looks like that an aggregate. This is also the case for the agglomerates of coarse RAP particles located in Regions R1, R3 and R6 for instance.

Previous observations suggest that a local stiffening of the binder occurs around certain RAP aggregates. This is attributed in this case to the presence of aged binder and fine RAP particles attached to the aggregates in these areas. In addition, the formation of agglomerations of small RAP particles is observed in some areas. The combination of these two phenomena may cause a modification of the mixture gradation leading to the changes in the grading curve design.

19.4 Conclusion

In the present study, the grid method was used to investigate the mechanical response in compression of asphalt mixtures incorporating various percentages of RAP. The mechanical response was observed at the binder scale. The strain fields obtained illustrate significant differences in the mechanical responses of the specimens under study. The 100 % RAP specimen exhibits highly concentrated strains around aggregates while more homogeneously distributed strain take place in the virgin mix. The analysis of the local behavior of 40 % RAP mixtures aggregates shows local stiffening of the binder in the vicinity of some recycled aggregates. As a general remark, the results obtained with this grid method are very promising to characterize the local behavior of RAP aggregates containing recycled mixtures.

Acknowledgment The Auvergne Regional Council, is gratefully acknowledged for providing the financial support of this study.

References

1. Zaumanis, M., Mallick, R.B., Frank, R.: 100% recycled hot mix asphalt: a review and analysis. Resour. Conserv. Recycl. **92**, 230–245 (2014)
2. Al-Qadi, I.L., Carpenter, S.H., Roberts, G., Ozer, H., Aurangzeb, Q., Elseifi, M., Trepanier, J.: Determination of Usable Residual Asphalt Binder in RAP. Illinois Center for Transportation (ICT), Rantoul (2009)
3. Grédiac, M., Toussaint, E.: Studying the mechanical behaviour of asphalt mixture with the grid method. Strain **49**, 1–15 (2013). doi:10.1111/str.12006
4. Grédiac, M., Toussaint, E., Petit, C., Millien, A., Ngyuen, D.C.: A comparative study of the heterogeneous local mechanical response of two types of asphalt mixes. Mater. Struct. **47**, 1513–1529 (2014). doi:10.1617/s11527-013-0132-3
5. Badulescu, C., Grédiac, M., Mathias, J.-D.: Investigation of the grid method for accurate in-plane strain measurement. Meas. Sci. Technol. **20**, 095102 (2009)
6. Piro, J.-L., Grédiac, M.: Producing and transferring low-spatial-frequency grids for measuring displacement fields with Moiré and grid methods. Exp. Tech. **28**, 23–26 (2004)
7. Grédiac, M., Sur, F.: Effect of sensor noise on the resolution and spatial resolution of displacement and strain maps estimated with the grid method. Strain **50**, 1–27 (2014)

Chapter 20
Assessment of Wood Properties Under Compression and Drying at the Ring Scale with the Grid Method

Djily Dang, Evelyne Toussaint, Rostand Moutou Pitti, and Michel Grédiac

Abstract The mechanical behavior, of Earlywood (EW) and Latewood (LW) of Abies Alba Mills (white fir of Massif Central) during drying under mechanical loading, is investigated. The grid method is used to measure the strain field and distinguish the heterogeneities at rings scale. Images are recorded by a Sensicam CCD camera. A 50-kN Zwick testing machine is used to apply the load. The tested specimen is first conditioned to a moisture content of 57.42 %. It is then tested and regularly removed in order to measure the moisture content. A tracking system ensuring constant positioning of the sample with respect to the camera has also been developed. Before loading, images have been taken to highlight the impact of moisture conditioning. Images are also grabbed during the loading steps and between them after totally unloading. All the strain maps are determined through a reference image (zero strain) taken before water conditioning of the specimen. During the experiments, strain gradients clearly appear between EW and LW, highlighting the variability of the mechanical properties at the rings scale. The effect of moisture content changes in the mechanical behavior of wood is also shown.

Keywords Drying • Early wood • Grid method • Late wood • Creep • Relaxation

20.1 Introduction

Wood is a particularly heterogeneous material with a complex behavior due to its composition, architecture, and hygroscopic nature [1]. Indeed, under the fiber saturation point (FSP, 25–30 %), any variation of its moisture content directly influences its mechanical and physical characteristics. It's also important to mention that, the effects of stress interfere with moisture variations and can increase deformations [2].

The behavior of Abies Alba Mills under creep and relaxation during drying is studied, using a full-field measurement technique named the grid method [3–5]. Such a measurement technique highlights, within certain limit, heterogeneities in strain fields that are caused here by the heterogeneous nature of the material.

The materials and methods are described in the first section of the paper. The results are then presented and discussed. A conclusion and some perspectives close the paper.

20.2 Materials and Methods

The goal is to study the deformation of wood sample during the creep and relaxation processes under natural drying. A suitable experimental setup, presented in Fig. 20.1, was designed for this purpose. The wood sample is Abies Alba Mills with the dimensions 47 × 43 × 17 mm^3 (see Fig. 20.2a). The investigated face of the specimen was equipped with a cross grid featuring a pitch equal to 200 μm (see Fig. 20.2b). The specimen was then immersed in water to reach a moisture content of 57.42 %. Grid images were recorded using a CCD camera at different times (every 15 s for the creep steps and every 5 min for the relaxation steps). The specimen was weighted at the beginning and at the end of every creep step. The same process was used for the relaxation phases. Figure 20.3 shows the curve of loading and unloading during the test.

D. Dang • E. Toussaint (✉) • R. Moutou Pitti • M. Grédiac
Institut Pascal, Clermont Université, Université Blaise Pascal, BP 10448, Clermont-Ferrand 63000, France

CNRS, UMR 6602, Institut Pascal, Aubière 63171, France
e-mail: evelyne.toussaint@univ-bpclermont.fr

Y. Zhu, A.T. Zehnder (eds.), *Experimental and Applied Mechanics, Volume 4*,
Conference Proceedings of the Society for Experimental Mechanics Series, DOI 10.1007/978-3-319-42028-8_20

Fig. 20.1 Experimental set-up

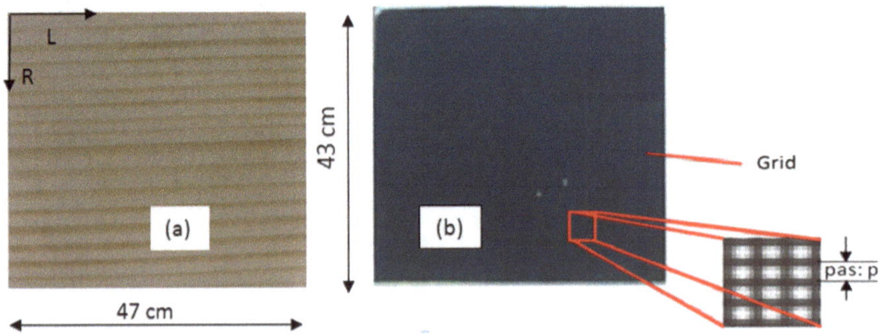

Fig. 20.2 Grid transfer procedure: (**a**) view of the face on which the grid is transferred, (**b**) transferred grid

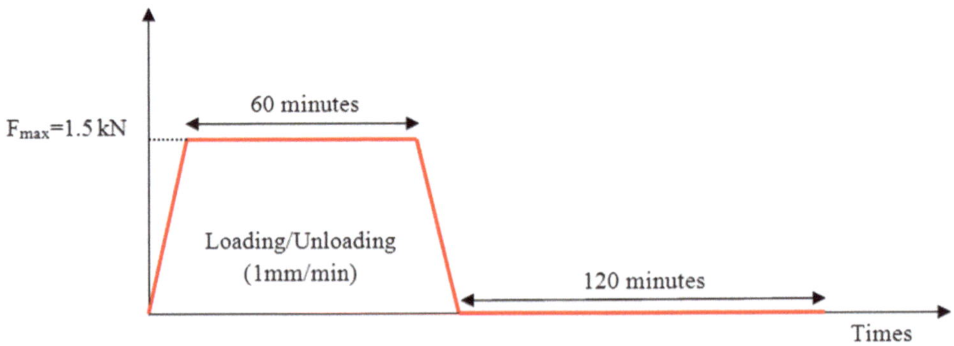

Fig. 20.3 Curve of loading and unloading during each cycle of creep and relaxation

20.3 Experimental Results and Discussion

The obtained results are presented below. Figure 20.4 shows the global moisture content evolution during the creep and the relaxation phases under drying. Moisture content is determined by weighting the specimen. As it can be seen, moisture content decreases nonlinearly during the natural drying process until a value of 11.64 %

- After the conditioning and before applying any mechanical load
  Strain maps calculated, after immersion and before mechanical loading, are presented in Fig. 20.5 (maps N° 2, 5 and 10). 10.7 h lasted between maps N°2 and N°10, during this time the global moisture content goes from 57.42 to 33.54 %. Deformation maps show that the specimen is under hydric expansion due to water adsorption. This swelling slightly persists after 10.7 h, even if the sample was removed from water immersion. One can note that the expansion strain is more pronounced in LW than in EW. This is agreement with the findings of Dang et al. [3, 4]. Indeed, the greater the density, the higher the hydric expansion.

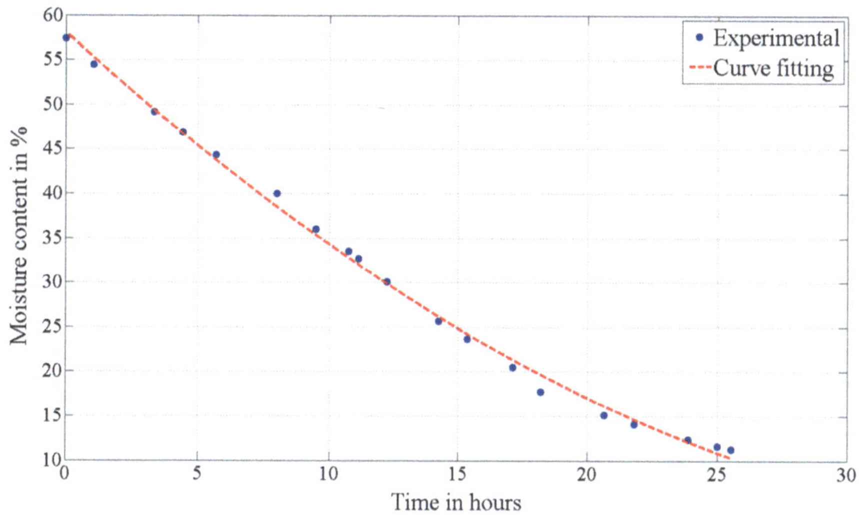

Fig. 20.4 Moisture content evolution

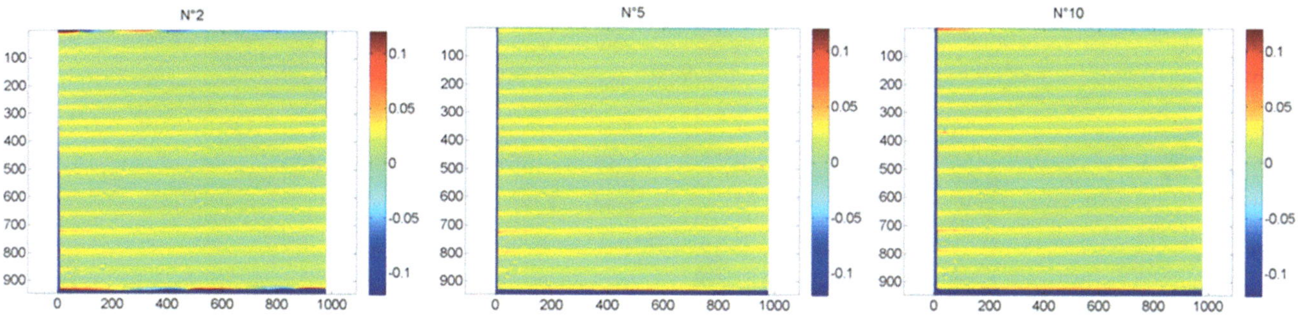

Fig. 20.5 Radial strain maps after immersion and before mechanical loading

- First loading (creep during drying)
  The first creep was performed 11 h after the immersion. During this phase, the global moisture content varies from 32.69 to 30.07 %. The strain maps at different steps during the first creep are presented in Fig. 20.6. In this latter, map N°2 shows the state of deformation at the beginning of the loading process. Map N°6 is obtained when the force nearly reaches the maximum value. Maps N°100 and 244 correspond to images registered after respectively 25 and 60 min of constant loading of $F_{max} = 1.5$ kN. We note that EW is submitted to a higher state of deformation due to its lower stiffness.
- Between the first and the second creep (relaxation during natural drying)
  The strain maps corresponding to the first relaxation are presented in Fig. 20.7. Maps N°2 and N°6 show the state of deformation after 5 and 30 min of relaxation, respectively. Maps N°15 and 25 are from images obtained after 75 and 125 min, respectively. Negative strains widely reduced during this first relaxation and the positive strain are still present mainly in LW. During this relaxation phase moisture content goes from 30.7 to 25.69 %.

  The same process was used to perform the second creep and relaxation at different ranges of moisture content.
- Creep and relaxation curves under natural drying
  The case A is obtained by grouping the first creep and relaxation, and the case B is the combination of the second creep and relaxation. Regions of interest (see Fig. 20.8) are chosen for each case to investigate the mean radial strain at the rings scale. From these chosen regions, masks are built to extract and to assess the mean radial deformation in EW and in LW, and for the whole defined region, during the creep and relaxation phases for each case.

  Figure 20.9 shows that the mean radial strains are higher (in absolute value) for case B than for case A; both in EW and in LW, and also in the delimited regions. This can be explained by the fact that, under the FSP, moisture content variations affect the mechanical properties of wood material. In addition, under stress and moisture content variations, wood is submitted to a supplementary strain; the mechano-sorptive one. Indeed, the total deformations investigated here include the elastic, viscoelastic, shrinkage and mechano-sorptive (during creep) effects. It can also be noted that deformations are higher in EW than in LW for both cases whatever the creep or the relaxation phases. This is due to the fact that LW is stiffer than EW.

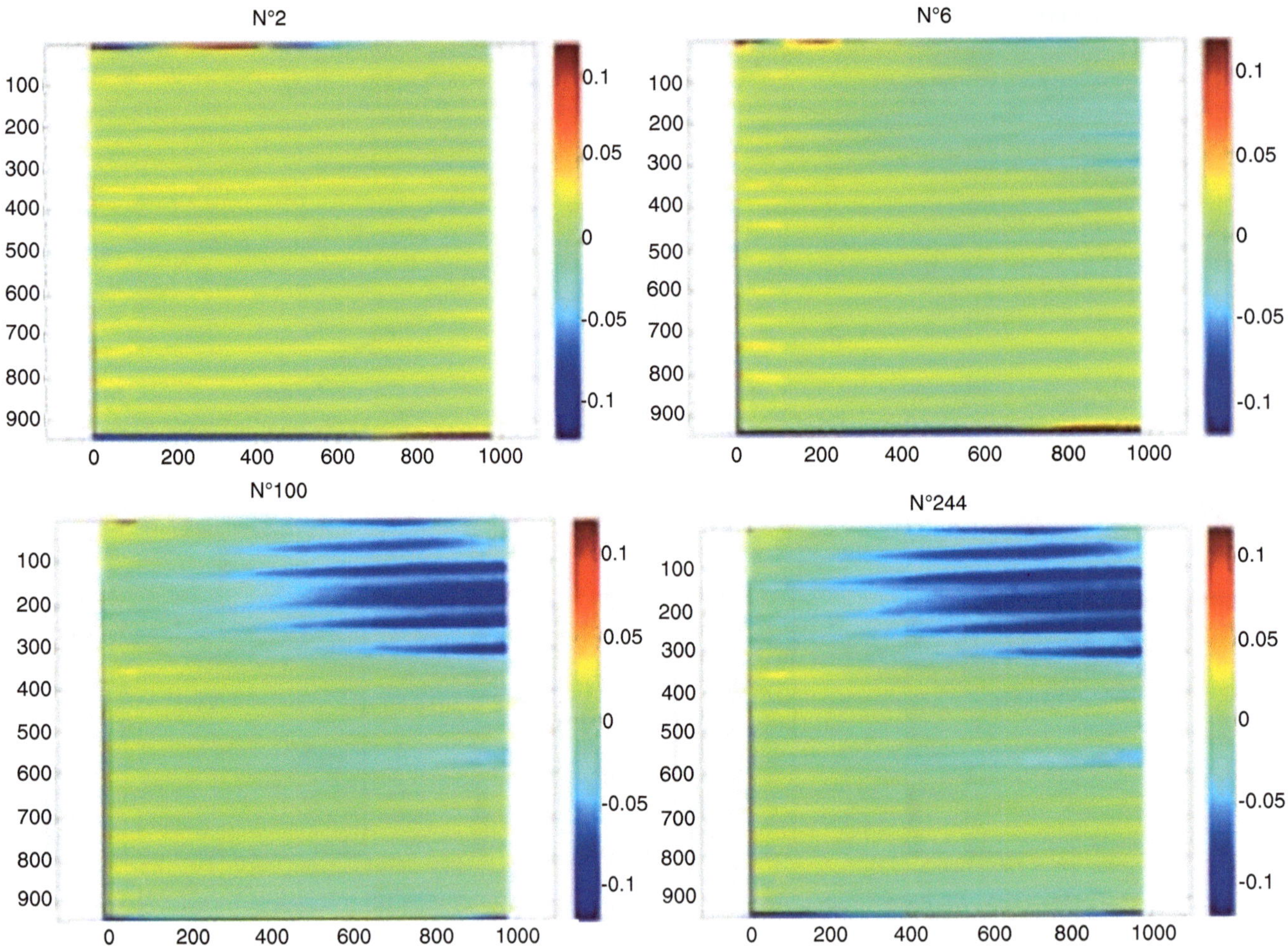

Fig. 20.6 Strain maps during the first creep

20.4 Conclusion

The phenomenon of creep and relaxation under natural drying was investigated for a specimen if Abies Alba Mills. For this purpose, a suitable experimental setup, was designed. The specimen was firstly immersed in water to reach a moisture content of 57.42 % before being submitted to a natural drying with cycles of creep and relaxation. Global moisture content was determined by weighting the specimen during the test.

Next step is to decouple the total deformation and to isolate each term in order to develop an inverse identification method to characterize wood under compressive stress and drying, at the growth rings scale.

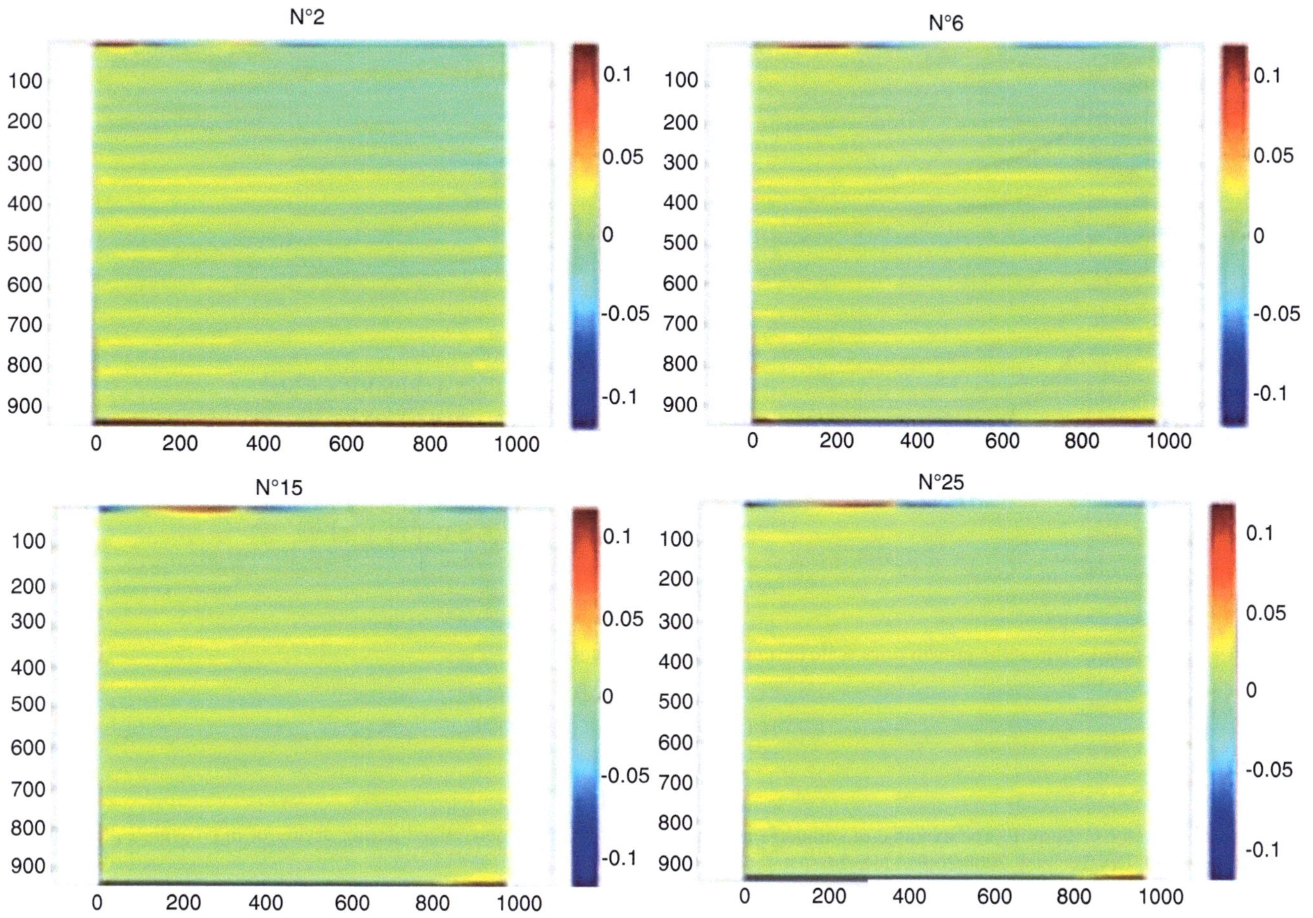

Fig. 20.7 Strain maps during the first relaxation

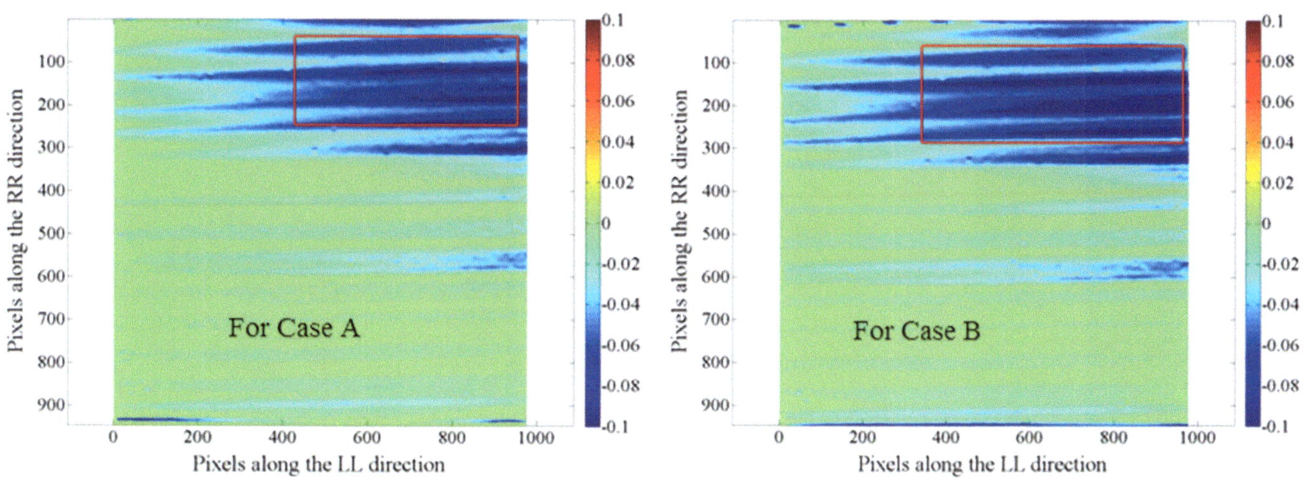

Fig. 20.8 Definition of zones of interest

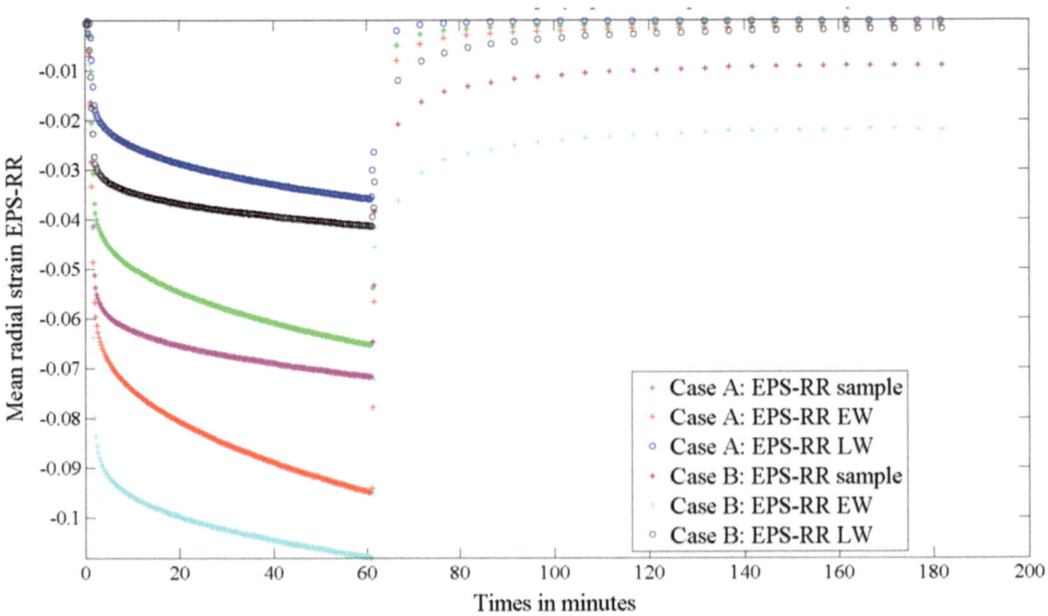

Fig. 20.9 Mean radial strain at the growth rings scale during creep and relaxation

References

1. Almeida G, Assor C, Perré P. The dynamic of shrinkage/moisture content behavior determined during drying of microsamples for different kinds of wood. Dry Technol. 2008;26(9):1118–24.
2. Hoffmeyer P, Davidson RW. Mechano-sorptive creep mechanism of wood in compression and bending. Wood Sci Technol. 1989;23 (3):215–27.
3. Dang, D., Moutou Pitti, R., Toussaint, E., Grédiac, M.: Experimental evidence of water diffusion gradient in wood using the grid method. In: Final Cost Action FP0904 Conference, Skellefteå, 19–21 May 2014
4. Dang D, Toussaint E, Moutou Pitti R, Grédiac M. Investigation of the hydric transfer phenomenon in wood at the ring scale with the grid method. In: Cesar S, John C, Paul G, editors. Experimental and Applied Mechanics, vol. 4. New York: Springer; 2016. p. 77–81.
5. Badulescu C, Grédiac M, Mathias J-D. Investigation of the grid method for accurate in-plane strain measurement. Meas Sci Technol. 2009;20 (9):1–17.

MIX
Papier aus verantwortungsvollen Quellen
Paper from responsible sources
FSC® C105338

If you have any concerns about our products,
you can contact us on
ProductSafety@springernature.com

In case Publisher is established outside the EU,
the EU authorized representative is:
Springer Nature Customer Service Center GmbH
Europaplatz 3, 69115 Heidelberg, Germany

Printed by Libri Plureos GmbH
in Hamburg, Germany